D1752036

Dror Sarid
Exploring Scanning Probe Microscopy with MATHEMATICA

1807–2007 Knowledge for Generations

Each generation has its unique needs and aspirations. When Charles Wiley first opened his small printing shop in lower Manhattan in 1807, it was a generation of boundless potential searching for an identity. And we were there, helping to define a new American literary tradition. Over half a century later, in the midst of the Second Industrial Revolution, it was a generation focused on building the future. Once again, we were there, supplying the critical scientific, technical, and engineering knowledge that helped frame the world. Throughout the 20th Century, and into the new millennium, nations began to reach out beyond their own borders and a new international community was born. Wiley was there, expanding its operations around the world to enable a global exchange of ideas, opinions, and know-how.

For 200 years, Wiley has been an integral part of each generation's journey, enabling the flow of information and understanding necessary to meet their needs and fulfill their aspirations. Today, bold new technologies are changing the way we live and learn. Wiley will be there, providing you the must-have knowledge you need to imagine new worlds, new possibilities, and new opportunities.

Generations come and go, but you can always count on Wiley to provide you the knowledge you need, when and where you need it!

William J. Pesce
President and Chief Executive Officer

Peter Booth Wiley
Chairman of the Board

Dror Sarid

Exploring Scanning Probe Microscopy with MATHEMATICA

Second, Completely Revised and Enlarged Edition

WILEY-VCH Verlag GmbH & Co. KGaA

The Author

Prof. Dror Sarid
College of Optical Sciences
University of Arizona
Tucson, Arizona 85721
USA
sarid@optics.arizona.edu

Cover:
Three buckyballs adsorbed on the surface of Si(100), obtained using ultra high vacuum STM.

■ All books published by **Wiley-VCH** are carefully produced. Nevertheless, authors, editors, and publisher do not warrant the information contained in these books, including this book, to be free of errors. Readers are advised to keep in mind that statements, data, illustrations, procedural details or other items may inadvertently be inaccurate.

Library of Congress Card No.:
applied for

British Library Cataloguing-in-Publication Data
A catalogue record for this book is available from the British Library

Bibliographic information published by the Deutsche Nationalbibliothek
The Deutsche Nationalbibliothek lists this publication in the Deutsche Nationalbibliografie; detailed bibliographic data are available in the Internet at <http://dnb.d-nb.de>.

© 2007 WILEY-VCH Verlag GmbH & Co. KGaA, Weinheim

All rights reserved (including those of translation into other languages). No part of this book may be reproduced in any form – photoprinting, microfilm, or any other means – transmitted or translated into a machine language without written permission from the publishers. Registered names, trademarks, etc. used in this book, even when not specifically marked as such, are not to be considered unprotected by law.

Typesetting Da-TeX Gerd Blumenstein, Leipzig
Printing Strauss GmbH, Mörlenbach
Binding Litges & Dopf Buchbinderei GmbH, Heppenheim

Printed in the Federal Republic of Germany
Printed on acid-free paper

ISBN: 978-3-527-40617-3

To Lea, Rami, Uri, Karen, and Danieli

Contents

Preface   17

1   **Introduction**   22
1.1   Style   22
1.2   *Mathematica* Preparation   23
1.2.1   General   23
1.2.2   Example   24
1.3   Recommended Books   25
1.3.1   Mathematica Programming Language   25
1.3.2   Scanning Probe Microscopies   26

2   **Uniform Cantilevers**   27
2.1   Introduction   27
2.2   Bending Due to F_z 30
2.2.1   General Equations   30
2.2.2   Slope   31
2.2.3   Angular Spring Constant   31
2.2.4   Displacement   32
2.2.5   Linear Spring Constant   32
2.2.6   Numerical Example: Si   33
2.2.7   Numerical Example: PtIr   33
2.3   Buckling Due to F_x 36
2.3.1   General Equations   36
2.3.2   Slope   36
2.3.3   Angular Spring Constant   36
2.3.4   Displacement   37
2.3.5   Linear Spring Constant   37
2.3.6   Numerical Example: Si   37
2.3.7   Numerical Example: PtIr   39
2.4   Twisting Due to F_y 40

2.4.1	General Equation	40
2.4.2	Slope	41
2.4.3	Angular Spring Constant	41
2.4.4	Numerical Example: Si	42
2.4.5	Numerical Example: PtIr	42
2.5	Vibrations	42
2.5.1	Bending Resonance Frequencies	42
2.5.2	Characteristic Functions	44
2.6	Summary of Results	45
	Exercises for Chapter 2	45
	References	46
3	**Cantilever Conversion Tables**	**48**
3.1	Introduction	48
3.2	Circular Cantilever	49
3.3	Square Cantilever	50
3.4	Rectangular Cantilever	52
	Exercises for Chapter 3	54
	References	55
4	**V-Shaped Cantilevers**	**56**
4.1	Introduction	56
4.2	Bending Due to F_z: Triangular Shape	58
4.2.1	General Equations	58
4.2.2	Slope	59
4.2.3	Angular Spring Constant	59
4.2.4	Displacement	59
4.2.5	Linear Spring Constant	60
4.2.6	Numerical Examples	60
4.3	Buckling due to F_x: Triangular Shape	62
4.3.1	General Equations	62
4.3.2	Slope	63
4.3.3	Angular Spring Constant	63
4.3.4	Displacement	63
4.3.5	Linear Spring Constant	64
4.3.6	Numerical Examples	64
4.4	Bending due to F_z: V Shape	66
4.4.1	General Equations	66
4.4.2	Slope	67
4.4.3	Angular Spring Constant	67
4.4.4	Displacement	67
4.4.5	Linear Spring Constant	68

4.4.6	Numerical Examples	68
4.5	Buckling Due to F_x: V Shape	70
4.5.1	General Equations	70
4.5.2	Slope	71
4.5.3	Angular Spring Constant	71
4.5.4	Displacement	71
4.5.5	Linear Spring Constant	72
4.5.6	Numerical Examples	72
4.6	Vibrations	74
4.6.1	Resonance Frequencies	74
4.6.2	Characteristic Functions	76
	Exercises for Chapter 4	77
	References	77
5	**Tip–Sample Adhesion**	**78**
5.1	Introduction	78
5.2	Indentation	81
5.2.1	Contact Radius and Contact Force	81
5.2.2	Indentation and Contact Radius	83
5.2.3	Indentation and Contact Force	84
5.3	Inverted Functions	85
5.3.1	Contact Force and Contact Radius	85
5.3.2	Contact Radius and Indentation	86
5.3.3	Contact Force and Indentation	86
5.3.4	Limits of Adhesion Parameters	87
5.4	Contact Pressure	88
5.4.1	Maximum Contact Pressure	89
5.4.2	Distribution of Contact Pressure	89
5.5	Lennard–Jones Potential	90
5.6	Total Force and Indentation	91
5.6.1	Push-in Region	91
5.6.2	Push-in Region in the Absence of Adhesion	91
5.6.3	Push-in Region in the Presence of Adhesion	92
5.6.4	Pull-out Region	93
5.6.5	Pull-out Region in the Absence of Adhesion	93
5.6.6	Pull-out Region in the Presence of Adhesion	93
5.6.7	Hysteresis Loop	93
	Exercises for Chapter 5	93
	References	94
6	**Tip–Sample Force Curve**	**95**
6.1	Introduction	95

6.2	Tip–Sample Interaction	97
6.2.1	Lennard–Jones Potential	97
6.2.2	Lennard–Jones Force	98
6.2.3	Lennard–Jones Force Derivative	99
6.2.4	Morse Potential	100
6.3	Hysteresis Loop	101
6.3.1	Snap-in and Snap-out Points	102
6.3.2	Calculated Hysteresis Loop	103
6.3.3	Observed Hysteresis Loop	103
6.4	Evaluation of Hamaker's Constant	107
6.5	Animation	108
	Exercises for Chapter 6	108
	References	108
7	**Free Vibrations**	**110**
7.1	Introduction	110
7.2	Equation of Motion	111
7.3	Analytical Solution	112
7.3.1	Equation of Motion	112
7.3.2	Steady-State Regime	113
7.3.3	Bimorph–Cantilever Phase	114
7.3.4	Q-Dependent Resonance Frequency	115
7.3.5	Frequency-Dependent Amplitude	116
7.3.6	Frequency at the Steepest Slope	117
7.3.7	Average Power	117
7.4	Numerical Solutions	117
7.4.1	Equation of Motion	117
7.4.2	Transient Regime	118
7.4.3	Bimorph–Cantilever Phase Diagram	118
7.4.4	Displacement–Velocity Phase Diagram	119
	Exercises for Chapter 7	119
	References	120
8	**Noncontact Mode**	**122**
8.1	Introduction	122
8.2	Tip–Sample Interaction	124
8.2.1	Lennard–Jones Potential	124
8.2.2	The Equation of Motion	126
8.2.3	Numerical Solution of the Equation of Motion	126
8.2.4	Approximate Analytical Solution of the Equation of Motion	129
	Exercises for Chapter 8	132
	References	132

9	**Tapping Mode** *133*
9.1	Introduction *133*
9.2	Lennard–Jones Potential *135*
9.3	Indentation Repulsive Force *136*
9.4	Total Tip–Sample Force *137*
9.5	General Solution *137*
9.6	Transient Regime *138*
9.7	Steady-State Regime *139*
9.8	Tapping Phase Diagram *140*
9.9	Displacement–Velocity Phase Diagram *140*
9.10	Numerical Value of the Phase Shift *140*
9.11	Summary of Results *144*
	Exercises for Chapter 9 *144*
	References *144*

10	**Metal–Insulator–Metal Tunneling** *146*
10.1	Introduction *146*
10.2	Tunneling Current Density *148*
10.2.1	General Solution *148*
10.2.2	Small Voltage Approximation *148*
10.2.3	Large Voltage Approximation *149*
10.3	The Image Potential *150*
10.4	Barrier with an Image Potential *152*
10.4.1	The Barrier *152*
10.4.2	The Barrier Width *153*
10.4.3	Average Barrier Height *154*
10.5	Comparison of the Barriers *154*
10.6	The General Solution with an Image Potential *156*
10.7	Apparent Barrier Height *157*
	Exercises for Chapter 10 *157*
	References *158*

11	**Fowler–Nordheim Tunneling** *160*
11.1	Introduction *160*
11.2	Fowler–Nordheim Current Density *161*
11.3	Numerical Example *163*
11.4	Oxide Field and Applied Field *164*
11.5	Oscillation Factor *165*
11.6	Averaged Oscillations *167*
11.7	Effective Tunneling Area *168*
	Exercises for Chapter 11 *169*
	References *169*

12	**Scanning Tunneling Spectroscopy** 170
12.1	Introduction 170
12.2	Fermi–Dirac Statistics 172
12.3	Feenstra's Parameter 173
12.4	Scanning Tunneling Spectroscopy 174
12.4.1	STS Data File 174
12.4.2	Data Processing 175
12.5	Spectroscopic Data 175
12.6	Comparison of STS Results 176
	Exercises for Chapter 12 177
	References 177
13	**Coulomb Blockade** 179
13.1	Introduction 179
13.2	Capacitance 180
13.2.1	Sphere–Plane Capacitance 181
13.2.2	Sphere–Sphere Capacitance 182
13.3	Quantum Considerations 182
13.4	Requirements and Approximations 184
13.5	Coulomb Blockade and Coulomb Staircase 184
13.5.1	Electrostatic Energy Due to the Charging of the Quantum Dot 185
13.5.2	Electrostatic Energy Due to the Applied Bias 185
13.5.3	Total Electrostatic Energy 185
13.6	Tunneling Rates 186
13.7	Tunneling Current 186
13.8	Examples 187
13.8.1	Example 1 187
13.8.2	Example 2 188
13.8.3	Example 3 189
13.8.4	Example 4 190
13.8.5	Example 5 191
13.9	Temperature Effects 192
13.9.1	Parameters 192
13.9.2	Very High Temperature Operation 193
13.9.3	Very Low Temperature Operation 193
13.9.4	Finite Temperature Operation 194
	Exercises for Chapter 13 196
	References 197
14	**Density of States** 198
14.1	Introduction 198
14.2	Sphere in Arbitrary Dimensions 199

14.3	Density of States in Arbitrary Dimensions	*201*
14.4	Density of States in Confined Structures	*203*
14.4.1	Quantum Wells	*203*
14.4.2	Quantum Wires	*204*
14.4.3	Cubical Quantum Dots	*204*
14.4.4	Spherical Quantum Dots	*205*
14.5	Interband Optical Transitions and Critical Points	*207*
	Exercises for Chapter 14	*208*
	References	*209*

15 Electrostatics *211*

15.1	Introduction	*211*
15.2	Isolated Point Charge	*212*
15.3	Point Charge and Plane	*212*
15.4	Point Charge and Sphere	*213*
15.5	Isolated Sphere	*214*
15.6	Sphere and Plane	*215*
15.6.1	Position of Charges Inside the Sphere	*215*
15.6.2	Magnitude of Charges Inside the Sphere	*216*
15.6.3	Position of Charges Outside the Sphere	*216*
15.6.4	Magnitude of Charges Outside the Sphere	*216*
15.6.5	Potential and Field	*217*
15.6.6	Potential Along the Axis of Symmetry	*217*
15.7	Capacitance	*217*
15.7.1	Sphere–Plane Capacitance	*218*
15.7.2	Example	*218*
15.8	Two Spheres	*218*
15.8.1	Capacitance: Exact Solution	*219*
15.8.2	Capacitance: Approximate Solution	*219*
15.8.3	Example	*219*
15.9	Electrostatic Force	*220*
	Exercises for Chapter 15	*220*
	References	*221*

16 Near-Field Optics *222*

16.1	Introduction	*222*
16.2	Far-Field Solution	*224*
16.2.1	Vector Potential	*224*
16.2.2	Electric Field	*224*
16.2.3	Electric Vector Field in the $\hat{x}$–$\hat{z}$ Plane	*225*
16.2.4	Electric Vector Field in the $\hat{x}$–$\hat{y}$ Plane	*225*
16.2.5	Magnetic Field	*226*

16.2.6	Poynting Vector and Intensity	227
16.2.7	Intensity in the $\hat{x}$–$\hat{z}$ Plane	228
16.2.8	Intensity in the $\hat{x}$–$\hat{y}$ Plane	229
16.3	Near-Field Solution	229
16.3.1	Transformation	229
16.3.2	Electric Field	230
16.3.3	Electric Field in the $\hat{x}$–$\hat{y}$ Plane	230
16.3.4	Magnetic Field	233
16.3.5	Poynting Vector and Intensity	233
16.4	Discussion of the Models	236
16.4.1	Electric Field	236
16.4.2	Intensity	237
16.5	Scattered Electric Fields Around Patterned Apertures	238
	Exercises for Chapter 16	238
	References	241
17	**Constriction and Boundary Resistance**	**242**
17.1	Introduction	242
17.2	A Metal as a Free Electron Gas	244
17.2.1	Lorenz Number, N_{Lorenz}, and the Wiedemann–Franz Law	244
17.2.2	Fermi Velocity, v_F	244
17.2.3	Fermi Temperature, T_F	245
17.2.4	Fermi k-Vector, k_F	246
17.2.5	Electron Density, N_e	246
17.2.6	Mean Free Path, ℓ	246
17.2.7	Ratio of ℓ/σ	247
17.2.8	Electronic Density of States, $\mathcal{D}_e$	247
17.2.9	Electronic Specific Heat, C_e	248
17.3	Constriction Resistance	248
17.3.1	Electrical Resistance in the Maxwell Limit	248
17.3.2	Electrical Resistance in the Sharvin Limit	251
17.3.3	Combined Electrical Resistance	254
17.4	Boundary Resistance	256
17.4.1	Thermal Boundary Resistance of General Media	256
17.4.2	Thermal Boundary Resistance of Metallic Media	258
17.4.3	Electrical Boundary Resistance of Metallic Media	260
	Exercises for Chapter 17	261
	References	261
18	**Scanning Thermal Conductivity Microscopy**	**263**
18.1	Introduction	263
18.2	Theory of Thermal Response	267

| 18.2.1 | Electrical and Thermal Circuits 267
| 18.2.2 | Cantilever Thermal Resistance and Temperature 268
| 18.2.3 | Tip–Sample Thermal Resistance 269
| 18.2.4 | Tip Thermal Resistance and Temperature 270
| 18.3 | Thermal and Mechanical Cantilever Bending 271
| 18.3.1 | Mechanical Bending 271
| 18.3.2 | Thermal Bending 272
| 18.3.3 | Combined Solution 273
| 18.4 | Results 275
| 18.4.1 | Tip-Side Coating, Upward Thermal Bending: Si and SiO_2 275
| 18.4.2 | Top-Side Coating, Downward Thermal Bending: Si and SiO_2 277
| 18.4.3 | Tip- and Top-Side Coating η-Dependent Apparent Height 279
| 18.4.4 | Tip- and Top-Side Coating κ-Dependent Apparent Height 279
| | Exercises for Chapter 18 280
| | References 281

19 Kelvin Probe Force Microscopy 282
19.1 Introduction 282
19.2 Capacitance Derivatives 285
19.2.1 Tip–Sample Capacitance Derivative 285
19.2.2 Cantilever–Sample Capacitance Derivative 287
19.3 Measurement of Contact Potential Difference 287
19.3.1 Tip–Sample and Cantilever–Sample Electrostatic Forces 287
19.3.2 Harmonic Expansion of Tip–Sample Force 289
19.3.3 Thermal Noise Limitations 291
Exercises for Chapter 19 291
References 292

20 Raman Scattering in Nanocrystals 293
20.1 Introduction 293
20.2 Raman Scattering in Bulk Silicon Crystals as a Function of Temperature 295
20.2.1 Introduction 295
20.2.2 Linewidth and Frequency Shift 295
20.2.3 Spectra 296
20.3 Raman Spectra in Nanocrystals at Room Temperature 298
20.3.1 Introduction 298
20.3.2 Linewidth and Frequency Shift 299
20.3.3 Spectra 299
20.4 Raman Spectra in Nanocrystals as a Function of Temperature 300
20.4.1 Introduction 300
20.4.2 Linewidth and Frequency Shift 301

20.4.3 Spectra *302*
Exercises for Chapter 20 *302*
References *303*

Index *305*

Preface

This second edition of the book *Exploring Scanning Probe Microscopy with Mathematica* is a revised and extended version of the first edition. It consists of a collection of self-contained, interactive, computational examples from the fields of scanning tunneling microscopy, scanning force microscopy, and related technologies, using *Mathematica* notebooks. It was written in *Mathematica* version 5.1 as a series of notebooks and was then translated into the TEX typesetting language. The software includes the code belonging to each chapter of the book. The files can be run independently of each other on any platform that supports *Mathematica* versions 5 and higher.

The main motivation for writing a book such as this arises from often-encountered situations where published models in the field of scanning probe microscopy require prior knowledge of other theoretical results. The reader of such material, therefore, needs to track down other publications that sometimes use different notations. A self-consistent, self-contained presentation would therefore be a real time-saver. A second motivation is the time-consuming effort required to code models that contain subtleties that are not easy to spot. The code presented in this book, being self-contained, alleviates this problem. A third motivation is associated with the benefit of working interactively with a live mathematical model and being able to change the values of its parameters. The computational results, which might range over unanticipated values, could provide better insight into the intricacies of a given problem than, say, reading plain text and browsing through several examples. The advantage of this book is that it provides an active approach to the study of and research in scanning probe microscopy.

This book can be used at several levels. At the first level, the reader can use the text, equations, figures, and examples for each case as one would with any other technical textbook. At a more advanced level, the reader who is familiar with the *Mathematica* programming language can download the code for each example from the attached CD and modify the different parameters to suit his particular needs. At the most advanced level, the reader can modify the programs by using more advanced theoretical treatments that either he or others in the field have developed.

This book consists of 20 chapters divided into five topics of interest in the field of scanning probe microscopy. (I) An introductory chapter containing technical discussions on how to run the code belonging to each chapter; (II) chapters dealing with atomic force microscopy that describe the mechanical properties of cantilevers, atomic force microscope tip–sample interactions, and cantilever vibration characteristics; (III) chapters on the theory and applications of tunneling phenomena consisting of metal–insulator–metal tunneling, Fowler–Nordheim field emission, scanning tunneling spectroscopy and Coulomb blockade; (IV) chapters on the density of states in arbitrary dimensions and electrostatics; (V) chapters dealing with near-field optics, scanning thermal conductivity microscopy, Kelvin probe microscopy, and anharmonic Raman scattering in nanocrystals.

All of the computer code presented in this book has already been used to model topics of interest at the Scanning Probe Microscopy Laboratory, College of Optical Sciences, University of Arizona. Because modeling of phenomena in atomic force microscopy, scanning tunneling microscopy, and related topics has been an ever-expanding activity, a large body of literature is available to the investigator. Nevertheless, some of these models require powerful computers or involve complicated code with elaborate theoretical considerations, neither of which are compatible with a book such as this. Also, models that do not belong to the mainstream of atomic force microscopy and scanning tunneling microscopy, or those whose range of validity has yet to be established, were not included in this book. It was decided that for this second edition only a small selected number of topics of high interest and wide applicability, whose coding is sufficiently simple, will be added to the first edition. Future editions will improve upon the topics discussed in this book and add new ones.

Most of the concepts presented in each chapter have already appeared in the literature either in detail or as brief comments. Some new insights, details, explanations, and examples, however, are introduced in practically each chapter of this book. These were made possible by the very fact that this book is about the mathematical modeling of the various topics, where numerical examples can be generated by the reader interactively.

The first chapter is an introduction that explains style conventions and presents a common list of units, and physical and material constants used in all the chapters of the book. Also included is the description of how the plots in this book have been produced. The second topic of the book treats atomic force microscopy in three chapters on cantilevers, two chapters on tip–sample interactions, and three chapters on modes of operation. Chapter 2, Uniform Cantilevers, presents the mechanical properties of uniform cantilevers having a solid, rectangular section. In particular, the bending and twisting of the cantilevers, and their resonance frequencies and characteristic functions are discussed. Chapter 3, Cantilever Conversion Tables, deals with uniform cantilevers having a rectangular or circular section. It makes possible to obtain one pair of the five parameters characterizing these cantilevers, such as length, radius or width, thickness, spring constant, and resonance frequency, in terms of the other three parameters. Chapter 4, V-Shaped Cantilevers, presents the linear and angular spring constants of these cantilevers, and their resonance frequencies and characteristic functions. Tip–Sample Adhesion is the topic of Chapter 5. Here, the interaction between the tip of an atomic force microscope and a sample in terms of a Johnson–Kendall–Roberts (JKR) adhesion model and a Lennard–Jones potential is described. The double-valued tip–sample contact force as a function of the indentation radius, with the resultant creation of a neck as the tip is pulled out of the sample, is also presented. Chapter 6, Tip–Sample Force Curves, treats the interaction between tip and sample as arising only from a Lennard–Jones potential, yielding a hysteresis loop on tip–sample approach and retraction. Chapter 7, Free Vibrations, models the cantilever as a driven, damped, linear oscillator, where the amplitude and phase of vibration are given as a function of the driving frequency and quality factor. Chapter 8, Noncontact Mode, describes the dependence of the resonance frequency, amplitude, and phase of vibration of a cantilever on the tip–sample force. It is shown that an approximate analytical solution involving the tip–sample force derivative yields an order of magnitude estimate for electric, magnetic, and atomic tip–sample forces. Chapter 9, Tapping Mode, presents the amplitude and phase of vibration of the cantilever and the tip–sample indentation force in terms of an attractive Lennard–Jones and repulsive indentation forces. As an example, the displacement, indentation, velocity, force, and pressure associated with tapping on soft and hard samples are presented.

The third topic of the book deals with scanning tunneling microscopy (STM). Metal–Insulator–Metal Tunneling is discussed in Chapter 10, where the basic principles of tunneling are presented together with a set of examples. The model considers a metal–insulator–metal (MIM) structure with two similar plane parallel metal electrodes that can be readily extended to dissimilar metals. A general tunneling equation is presented, and approximate solutions that include the image potential for small and large voltages are given. Chapter 11, Fowler–Nordheim Tunneling, describes the field emission, or tunneling, of electrons through a metal–oxide–semiconductor (MOS) structure. Also presented are the oscillations in the tunneling current due to resonance effects of electrons traveling in the conduction band of the oxide. Scanning Tunneling Spectroscopy is the topic of Chapter 12, presenting a code that can be used to process a scanning tunneling microscope current against voltage data and generate plots of $i(v)$, $\partial i/\partial v$, and the logarithmic derivative $\partial \ln i/\partial \ln v$. Ultra-high vacuum scanning tunneling microscopy of C_{60} molecules chemisorbed on a Si(100)–2 × 2 surface is used as an example to illustrate the power of this spectroscopic technique. Chapter 13, Coulomb Blockade, describes the principles of single-electron transistors, using an approximate model that replicates a tunneling current against the applied voltage of five experimental cases cited in the literature.

The fourth topic of the book consists of three chapters describing phenomena encountered in scanning probe microscopy where structures on the nanometer scale are being fabricated and characterized. Chapter 14, Density of States, presents the density of electronic states of large bodies in arbitrary dimensions, and quantum wells, quantum wires, and cubic and spherical quantum dots. Chapter 15, Electrostatics, presents exact and approximate sphere–plane and sphere–sphere capacitances together with the electrostatic force between a conducting sphere and a conducting plane, applicable to situations where the probing tip is a conductor. Chapter 16, Near-Field Optics, discusses the Bethe–Bouwkamp solution to light diffracted by a circular aperture and presents plots of electric fields and intensities in the near and far field. Chapter 17 to Chapter 20 have been added to the first edition.

Chapter 17, Constriction and Boundary Resistance, summarizes recent results of thermal and electrical resistances due to (a) constriction of flow of phonons and electrons through a narrow aperture, and (b) boundaries between two media. Chapter 18, Scanning Thermal Conductivity Microscopy, describes thermal flow from a laser-heated cantilever into two paths; in one path the flow through the tip and into the sample is controlled by the thermal conductivities of the tip–sample interface and the sample, and in the other path the flow is toward the base of the cantilever. By using a metal-coated cantilever one can get a map of the thermal conductivity across a sample from the thermal bending of the coated cantilever. Chapter 19, Kelvin Probe Microscopy, describes the operation of a cantilever driven by the application of a tip–sample voltage at a frequency ω in the presence of a dc tip–sample contact potential difference (CPD). While the ac voltage drives the cantilever at 2ω, the CPD gives rise to a vibration at ω. The external dc voltage between the tip and the sample, required to null the vibration at ω, is a measure of the CPD. Chapter 20 describes anharmonic Raman scattering in nanocrystals, a topic of much interest lately where a metal-coated tip of an atomic force microscope is used to generate local Raman enhancement by several orders of magnitude.

To use the book efficiently, the reader should read the papers cited at the end of each chapter. They provide a short introduction to the subject matter, present the main ideas, and offer references to relevant topics. Excellent books on the field of scanning tunneling microscopy, atomic force microscopy, and related topics are readily available. They can serve as powerful tools in using the different topics discussed in this book. Comments from the users of the material presented in this book will be invaluable in making a second edition more accurate, efficient, and useful.

The two editions of this book took about 3 years to write, but the ideas, the methods used to present them, and the implementation of the code and its testing took many more years. All of this work was made possible by Gerd Binnig and Heinrich Rohrer, the fathers of scanning tunneling microscopy, and Gerd Bininig, Calvin Quate, and Christopher Gerber, the fathers of atomic force microscopy. The able help, inspiration, and encouragement received during this period of time from Todd G. Ruskell, Richard K. Workman, Xiaowei Yao, Charles A. Peterson, Jeffery P. Hunt, Guanming Lai, Robert D. Grober, Dong Chen, Ralph Richard, Brendan Mc Carthy, Ranjan Grover, and Pramod Khulbe were indispensable indeed. The work on the two editions of this book was kindly supported by partial funding from the National Science Foundation, Office of Naval Research, Ballistic Missile Defense Office, National Aeronautics and Space Administration, Department of Energy, National Institute of Science and Technology, and IBM, Veeco, EMC, Motorola, and NanoChip corporations.

Special thanks are also due to the editors of Wiley and to Margaret Regan for their able editing effort.

Tucson, August 2006
Dror Sarid

1
Introduction

1.1
Style

The style of the first edition of this book, which carries the same title as this second edition, consisted of a mixture of TEX-like equations and equations generated by *Mathematica* computational output, interwoven between each other. There were two reasons for this choice of style. The first reason is that at the time when the first edition was written, the newest version of *Mathematica* was version 2.2, which was limited in its ability to produce as an output the traditional form of equations. It was found necessary, therefore, to print most of the equations using TEX. This new edition of the book is using *Mathematica* version 5.2, whose text form is close enough to TEX to make the equations appear similar to the traditional form. The second, and more important reason for having equations printed in a combination of TEX and *Mathematica* computational output was the belief that the reader will benefit from having the code running the simulations in each chapter transparent to him. Consequently, the text was interwoven with the segments of the chapter's code, making it easy to modify each segment "on the run."

After 6 years of using this book as both a research and a teaching resource, it became apparent that the code and the text should be completely separated. There were two reasons for the need of this separation. The first reason stems from the fact that expertise in both *Mathematica* and scanning probe microcopy (SPM) is not as prevalent among the SPM community as originally thought. The second and more important reason for the need of this separation is based on the fact that the styles required for coding and composing text are very different. It is much easier to compose the code that runs a chapter without having to pay attention to the demands required by composing a text. In contrast, it is much easier to compose the text without having to be limited by the shortcomings of the style of the *Mathematica* computational output.

This second edition of the book has 20 chapters, each of which consists of three components. The first component, titled *Mathematica* Preparation, is a code that is common to all the chapters in this book. This code, as will be described in detail, is a collection of bits of information needed by the specific code belonging to all the chapters. This code is attached to the code associated with each of the chapters. The second component is the code that is specific to each chapter. This code generates the tables and figures appearing in the chapters using typical parameters. These parameters can be changed within a reasonable range, generating new tables and figures. Although there is no text embedded in the code, it is clear what is the function of each *Mathematica* instruction it contained when read together with its associated text. The third component consists of the printed text of the chapters of the book that contains no code at all. The equations appearing in this component are renditions of the *Mathematica* code that were rearranged to appear close to the TeX form.

By dividing each chapter into these three components, one gains several advantages that were proven time and again to be extremely useful in both research and teaching. Having the first component shared by all the chapters insures a common style to the parameters, tables, and figures. Having the code of each chapter separated from its text makes it easier to develop research ideas and test them based only on the merit of the results presented by table and figures. Following this method requires no attention to a clear presentation of computational results. After the code is developed, one can compose a code-free text with traditionally recognized equations, making the presentation of scientific arguments clear and simple.

1.2
Mathematica Preparation

1.2.1
General

The first step in running the code belonging to each chapter consists of **Clearing** previous computational results and turning off comments addressing usage of similar names for different routines. The code belonging to each chapter may require the use of several **Packages,** all of which will therefore be loaded. The code uses a standard notation of **Units** for numerical calculations. To facilitate algebraic solutions, we use a numerical subroutine, **NSub**, as a replacement rule for all the **Physical constants** used in the chapters. Thus, one can develop a model of a physical problem and obtain algebraic results. By using the replacement rule, the results are rendered numerically. A collection of material constants used or can be used in the code includes **Young's modulus**, E, **Poisson's ratio**, ν, **Mass density**, ρ, **Electric conductivity and resistivity**, σ_e and ρ_e, respectively, **Thermal conductivity**, κ, and **Thermal expansion**, α. To each of these the designated material name is appended. To shorten the code of the figures in the book, we use a **Plotting Style** and options that contain the most common plotting commands. These options include **General option, Option for solid lines, Option for dashed lines**, and **Simple option**. The *Mathematica* **Preparation** code is included in the code of each chapter.

1.2.2
Example

Figure 1.1 is an example of plots of three functions, $\sin x$, $\sin 2x$ and $\sin 3x$, using the option **opt1**. The code sets the minimum and maximum values of the horizontal and vertical axes, has a frame label that reads a given parameter, $a = 2.3456$, for example, and uses **Epilog** to have text embedded in the figure that reads a given parameter. The presented values of the parameters can also be controlled by **IntegerPart** .

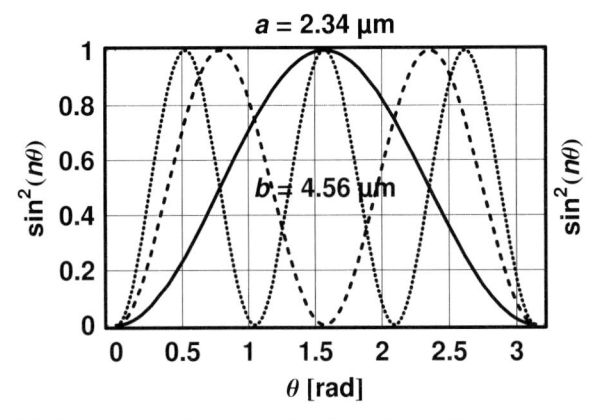

Fig. 1.1 An example of a figure using the option **opt1**.

1.3 Recommended Books

1.3.1 Mathematica Programming Language

The following is a selected list of books on the *Mathematica* programming language that can be used both as a teaching and a refresher tool.

1. Stephen Wolfram, *Mathematica, A System for Doing Mathematics by Computer*, 2nd Edition, Addison-Wesley, Reading, MA, 1991.

2. R. E. Crandall, *Mathematica for the Sciences*, Addison-Wesley, Reading, MA, 1991.

3. N. Blachman, *Mathematica: A Practical Approach*, Prentice Hall Series in Innovative Technology, Prentice Hall, Englewood Cliffs, NJ, 1992.

4. W. T. Shaw and J. Tigg, *Applied Mathematica: Getting Started, Getting It Done*, Addison-Wesley, Reading, MA, 1994.

5. S. Kaufmann, *Mathematica as a Tool: An Introduction with Practical Examples*, Birkhauser, Basel, 1994.

6. T. B. Bahder, *Mathematica for Scientists and Engineers*, Addison-Wesley, Reading, MA, 1995.

7. P. P. Tam, *A Physicist's Guide to Mathematica*, Academic Press, New York, 1997.

8. B. F. Torrence and E. A. Torrence, *The Student's Introduction to Mathematica: A Handbook for Precalculus, Calculus, and Linear Algebra*, Cambridge University Press, Cambridge, 1999.

9. S. Wagon, *Mathematica in Action*, Springer, Telos, 2000.

10. M. H. Hoft and H. F. W. Hoft, *Computing with Mathematica*, Academic Press, New York, 2002.

11. S. Wolfram, *Mathematica Book*, 5th Edition, Cambridge University Press, Cambridge, 2003.

12. C. -K. Cheung, et al., *Getting Started with Mathematica*, 2nd Edition, Wiley, New York, 2003.

13. G. Baumann, *Mathematica for Theoretical Physics: Classical Mechanics and Nonlinear Dynamics*, Springer, New York, 2005.

14. M. Trott, *The Mathematica Guidebook for Programming*, Springer, Berlin, 2004.

1.3.2
Scanning Probe Microscopies

The literature on scanning probe microscopies, which grew almost exponentially in the past decade, includes papers, review articles, and books. Of these we selected a list of those books that cover the topics dealt with in this book.

1. R. J. Behm, N. Garcia, and H. Rohrer, eds., *Scanning Tunneling Microscopy and Related Methods*, NATO ASI Series E 184, Kluwer, Dordrecht, 1990.
2. D. Sarid, *Scanning Force Microscopy with Applications to Electric, Magnetic, and Atomic Forces*, Oxford University Press, New York, 1991.
3. H.-J. Guntherodt and R. Wiesendanger, eds., *Scanning Tunneling Microscopy I*, Springer Series in Surface Sciences 20, Springer, New York, 1992.
4. R. Wiesendanger and H. J. Guntherodt, eds., *Scanning Tunneling Microscopy II*, Springer Series in Surface Sciences 28, Springer, New York, 1992.
5. R. Wiesendanger and H. J. Guntherodt, eds., *Scanning Tunneling Microscopy III*, Springer Series in Surface Sciences 29, Springer, New York, 1993.
6. Ph. Avouris, ed., *Atomic and Nanometer-Scale Modification of Materials: Fundamentals and Applications*, NATO ASI Series E 239, Kluwer, Dordrecht, 1993.
7. C. Julian Chen, *Introduction to Scanning Tunneling Microscopy*, Oxford University Press, New York, 1993.
8. D. W. Pohl and D. Courjon, eds., *Near Field Optics*, NATO ASI Series E 242, Kluwer, Dordrecht, 1993.
9. D. Sarid, *Scanning Force Microscopy with Applications to Electric, Magnetic, and Atomic Forces*, Revised Edition, Oxford University Press, New York, 1994.
10. R. Wiesendanger, *Scanning Probe Microscopy and Spectroscopy, Methods and Application*, Cambridge University Press, Cambridge, 1994.
11. Y. Martin, ed., *Scanning Probe Microscopes, Design and Applications*, SPIE Milestone Series, Volume MS 107, 1995.
12. M. A. Paesler and P. J. Moyer, *Near-Field Optics*, Wiley, New York, 1996.
13. S. N. Magonov and M.H. Whangbo, *Surface Analysis with STM and AFM, Experimental and Theoretical Aspects of Image Analysis*, VCH, New York, 1996.
14. D. Sarid, *Exploring Scanning Probe Microscopy with Mathematica*, Wiley, NY, 1997.
15. C. Bai, *Scanning Tunneling Microscopy and Its Applications*, New York, Springer, 2000.
16. S. Morita, R. Wiesendanger, and E. Meyer, *Noncontact Atomic Force Microscopy*, Springer, New York, 2002.
17. Ernst Meyer, Hans J. Hug, and Roland Bennewitz, *Scanning Probe Microscopy: The Lab on a Tip*, Springer, New York, 2004.

2
Uniform Cantilevers

■ Highlights

1. Effects of bending, buckling, and twisting
2. Displacement and slope
3. Linear and angular spring constants
4. Resonance frequencies and characteristic functions

Abstract

The mechanical properties of cantilevers used in atomic force microscopy (AFM) play a key role in determining their ability to perform specific tasks. In particular, the bending, buckling, and twisting of the cantilevers, and their resonance frequencies and characteristic functions have to be optimized for samples with a given stiffness and for different scanning speeds. This chapter treats uniform cantilevers having a rectangular section and presents their properties both algebraically and numerically.

2.1
Introduction

Figure 2.1 shows the geometry of a typical uniform, rectangular-section cantilever used in atomic force microscopy (AFM), and a sample that is in contact with the sharp apex of the tip fabricated close to the free end of the cantilever. The cantilever body is positioned along the $\hat{x}$ direction, its tip, located at a distance δ from the free end of the cantilever, points in the $-\hat{z}$ direction, and the surface of the sample is in the $\hat{x}$–$\hat{y}$ plane. The cantilever length, width, and thickness, and the tip height are denoted by L, w, d, and h_t, respectively. In the following, the subscripts associated with the various parameters and functions denote the directions $\hat{x}$, $\hat{y}$, and $\hat{z}$ along which particular forces act. For the calculations presented in this chapter, we will ignore δ since it is much smaller than L. The cantilever is usually mounted at an angle of approximately 12.5° relative to the sample, to prevent contact between its base and elevated structures across the surface of the sample. For simplicity, this angle will be ignored in future calculations without any meaningful loss of accuracy. When the AFM operates in the contact mode, the cantilever is lowered toward the surface of the sample until the tip–sample force that acts in the $\hat{z}$ direction, F_z, reaches a prescribed value determined by the cantilever bending.

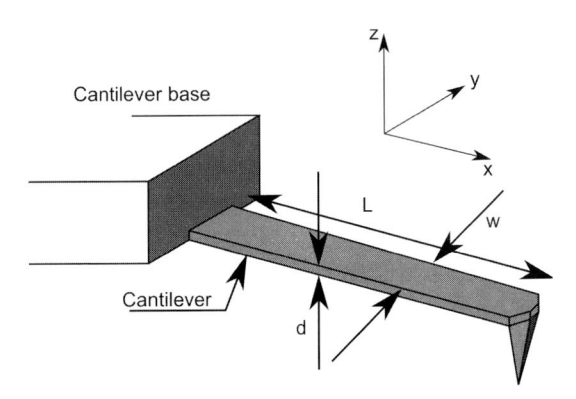

Fig. 2.1 A schematic diagram of a uniform, rectangular-section cantilever and its tip.

During the raster-scanning process of the AFM, the tip of the cantilever is dragged across the surface of the sample in the $\hat{x}$ or $\hat{y}$ direction. There are two forces acting on the tip when it is scanned along either direction. The first force, F_z, that acts in the $\hat{z}$ direction is common to both scan directions. It is associated with the rise and fall of the tip as it scans along topographic features. This force will give rise to a bending of the cantilever. For scanning in the $\hat{x}$ direction, there is also a tip–sample friction force, F_x, that acts in the $\hat{x}$ direction, which buckles the cantilever. When the tip is scanned in the $\hat{y}$ direction, it encounters not only the topographically related force, F_z, but also a friction force, F_y, that acts in the $\hat{y}$ direction and twists the cantilever. Under the bending and buckling actions, the cantilever develops a curvature across its body that is directed along the $\hat{y}$ axis, whose slope is denoted by $\theta_z(x)$ and $\theta_x(x)$, respectively. The bending and buckling of the cantilever also give rise to a displacement along its body, $\delta_z(x)$ and $\delta_x(x)$, both in the $\hat{z}$ direction. Under a twisting action, the curvature of the cantilever is along the $\hat{x}$ axis with a slope denoted by $\phi_y(x)$.

The AFM is using a laser diode to monitor the deflection of the cantilever during the scanning process. The beam of the laser is incident on the cantilever from which it is reflected into a quadrant photodiode. The resultant photocurrents, which depend on the slope of the cantilever at the point of incidence, are monitored and processed by a computer. The pair of quadrants positioned along the $\hat{z}$ axis respond to the bending and buckling slopes of the cantilever, while those positioned along the $\hat{y}$ axis respond to the twisting slope. The AFM can distinguish between the bending- and buckling-induced slopes of the cantilever by performing a scan in the $\hat{x}$ direction followed by a scan in the $-\hat{x}$ direction. These two scan directions produce different apparent topographic images because the friction force in each case will either be added to or subtracted from the true topographic contribution. The topographic images obtained by scanning in the $\hat{x}$ and $\hat{y}$ directions, however, are generated by the quadrant pairs positioned along the $\hat{x}$ and $\hat{y}$ axes, respectively, and can therefore be independently interpreted.

It is important to recognize that the cantilever interacts with the surface of the sample by forces acting on its tip which is located at a distance δ from its free end. The laser beam, in contrast, is incident at a particular point across the cantilever, not necessarily at its free end. The position of the laser spot on the cantilever depends on the particular AFM model and cantilever make. Therefore, for a proper interpretation of the bending of a given cantilever, one has to know both its deflection, generated by the tip–sample forces, and its slope at a particular point from which the beam of the laser is deflected.

In the following we treat the bending, buckling, and twisting of the cantilever when the AFM operates in the contact mode, and the resonance frequencies of the cantilever with their associated characteristic functions when the AFM operates in the noncontact mode.

The theory presented in this chapter is accompanied by numerical examples that relate to two types of cantilevers. For all the examples presented in this chapter, the figures describe plots associated with $F_i = 10$ (solid line), 20 (dashed line), and 30 nN (dotted line), where $i = x, y,$ or z. The first example uses the parameters of an all-silicon, uniform, rectangular-section cantilever having a sharp tip fabricated at its free end. The parameters of this cantilever are given in Table 2.1.

Table 2.1 The parameters of an all-silicon, uniform, rectangular-section cantilever having a sharp tip fabricated at its free end.

$$L = 0.2\,\text{mm}$$
$$w = 50\,\mu\text{m}$$
$$d = 1\,\mu\text{m}$$
$$h_t = 5\,\mu\text{m}$$
$$\delta = 5\,\mu\text{m}$$
$$E = 179\,\text{GN}/\text{m}^2$$
$$\rho = 2330\,\text{kg}/\text{m}^3$$

The second, less-used, yet important cantilever is an all-metal, uniform, rectangular-section cantilever fabricated from a flattened wire made of a platinum–iridium alloy. One end of such a hand-made cantilever is soldered to a small metal base, and the other is chemically etched and bent to form a sharp tip. Although such a cantilever is hard to mass-produce, it excels in applications where the local electric conductivity of a sample is probed, because it is robust, resists ablation, and remains conducting under ambient conditions. The parameters of this cantilever are given in Table 2.2.

Table 2.2 The parameters of a uniform, rectangular-section cantilever made of PtIr having a sharp tip fabricated at its free end.

$$L = 2.5\,\text{mm}$$
$$w = 100\,\mu\text{m}$$
$$d = 20\,\mu\text{m}$$
$$h_t = 500\,\mu\text{m}$$
$$\delta = 10\,\mu\text{m}$$
$$E = 240\,\text{GN/m}^2$$
$$\rho = 21\,620\,\text{kg/m}^3$$

2.2
Bending Due to F_z

2.2.1
General Equations

Consider a cantilever that scans a sample in an arbitrary direction in the $\hat{x}$–$\hat{y}$ plane, where the tip–sample force, F_z, generates a displacement, $\delta_z(x)$. The slope of the cantilever, R, in terms of its Young's modulus E, its area moment of inertia, I_z, and the bending moment, $M_z(x)$, is given by

$$\frac{1}{R} = \frac{M_z}{EI_z}. \tag{2.1}$$

On the other hand, the curvature, R, of a general function and in particular of $\delta_z(x)$, is given by

$$\frac{1}{R} = \frac{\frac{\partial^2}{\partial x^2}\delta_z(x)}{\left[1 + \left(\frac{\partial}{\partial x}\delta_z(x)\right)^2\right]^{3/2}}. \tag{2.2}$$

Equating the two expressions for R and assuming a small curvature, leads to the equation of bending of the cantilever,

$$\frac{\partial^2}{\partial x^2}\delta_z(x) = \frac{M_z(x)}{EI_z}. \tag{2.3}$$

The bending moment, M_z, associated with the neutral axis of the cantilever, which is induced by the force F_z, is

$$M_z(x) = (L - x)F_z. \tag{2.4}$$

The axial moment of inertia of the cantilever, I_z, associated with the axis that passes along its centroid, is

$$I_z = \int_A z^2 \, dA = \int_{-d/2}^{d/2} \int_{-w/2}^{w/2} z^2 \, dx \, dz, \tag{2.5}$$

where $dA = dx\,dz$ is an element of area. The integration in Eq. (2.5) yields

$$I_z = \frac{wd^3}{12}. \tag{2.6}$$

Inserting Eq. (2.4) and Eq. (2.6) into Eq. (2.3) gives an explicit, second-order differential equation that relates the displacement of the cantilever to the bending force,

$$\frac{\partial^2}{\partial x^2} \delta_z(x) = 12 \frac{L-x}{Ewd^3} F_z. \tag{2.7}$$

Equation (2.7) will form the basis to the modeling of the response of the cantilever to F_z.

2.2.2
Slope

The slope of the cantilever at a point x along its length, $\theta_z(x)$, is obtained by integrating Eq. (2.7),

$$\theta_z(x) = 6 \frac{(2L-x)x}{Ewd^3} F_z, \tag{2.8}$$

which, at the free end of the cantilever, where $x = L$, gives

$$\theta_z(L) = 6 \frac{L^2}{Ewd^3} F_z. \tag{2.9}$$

2.2.3
Angular Spring Constant

We define the angular spring constant of the cantilever, $k_{\theta_z}(x)$, as the ratio of the force acting on its tip and the resultant slope of the cantilever at the point x,

$$k_{\theta_z}(x) = \left| \frac{F_z}{\theta_z(x)} \right|. \tag{2.10}$$

Inserting Eq. (2.9) into Eq. (2.10) yields

$$k_{\theta_z}(x) = \frac{Ewd^3}{6(2L-x)x}. \tag{2.11}$$

For $x = L$, one obtains a particular value of the angular spring constant,

$$k_{\theta_z} = \frac{Ewd^3}{6L^2}. \tag{2.12}$$

2.2.4
Displacement

The local displacement of the cantilever in the $\hat{z}$ direction, $\delta_z(x)$, is obtained by integrating the slope, Eq. (2.8), which yields

$$\delta_z(x) = \frac{2x^2(3L - x)}{Ewd^3} F_z. \tag{2.13}$$

At the free end of the cantilever, where $x = L$, the slope is

$$\delta_z(L) = \frac{4L^3}{Ewd^3} F_z. \tag{2.14}$$

2.2.5
Linear Spring Constant

The cantilever linear spring constant, k_z, is defined as the ratio of the force acting on its tip to the resultant displacement at $x = L$,

$$k_z = \left|\frac{F_z}{\delta_z(L)}\right|. \tag{2.15}$$

The linear spring constant of the cantilever determines (a) its displacement for a given tip–sample force and (b) its resonance frequency, as will be discussed later. Inserting Eq. (2.13) into Eq. (2.15) yields

$$k_z = \frac{Ewd^3}{4L^3}. \tag{2.16}$$

Note the strong dependence of k_z on the thickness, d, and the length, L, of the cantilever. Remember that tip–sample forces always act on the tip of the cantilever, while the beam of the laser is incident at some point x along its body. It is therefore of interest to examine the ratio of the angular and linear spring constants, given by the function $\tilde{\zeta}_z(x)$,

$$\tilde{\zeta}_z(x) = \frac{k_{\theta_z}(x)}{k_z}. \tag{2.17}$$

Inserting Eq. (2.11) and Eq. (2.16) into Eq. (2.17) gives

$$\tilde{\zeta}_z(x) = \frac{3(2L - x)x}{2L^3}. \tag{2.18}$$

The function $\zeta_z(x)$ is used by the software of the AFM to calibrate the tip–sample force using the specifications of the particular cantilever in use.

2.2.6
Numerical Example: Si

We can now present numerical examples for the response of the Si cantilever to a force acting in the $\hat{z}$ direction. Figure 2.2 shows the slope of the cantilever, $\theta_z(x)$, as a function of the distance x from its base for $F_z = 10$ (solid line), 20 (dashed line), and 30 nN (dotted line). The angular spring constant of the cantilever, k_{θ_z}, is depicted in the figure. Note that the slope is the steepest next to the base of the cantilever, where it is most likely to break.

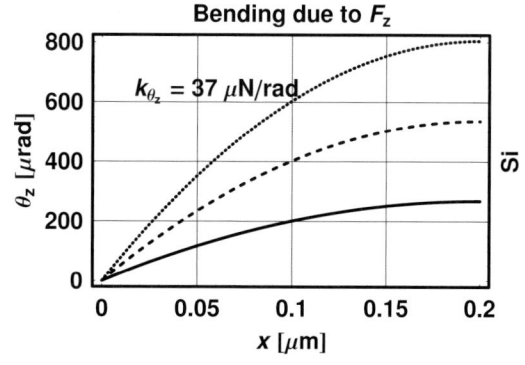

Fig. 2.2 The slope of the Si cantilever, $\theta_z(x)$, as a function of the distance x from its base, for a tip–sample force acting in the $\hat{z}$ direction for $F_z = 10$ (solid line), 20 (dashed line), and 30 nN (dotted line).

Figure 2.3 shows the displacement of the Si cantilever, $\delta_z(x)$ as a function of the distance x from its base for $F_z = 10$ (solid line), 20 (dashed line), and 30 nN (dotted line). Here, the linear spring constant of the cantilever, k_z, is depicted in the figure.

Figure 2.4 shows the ratio of the angular and linear spring constants of the Si cantilever, $\zeta_z(x)$, as a function of the distance x from its base.

2.2.7
Numerical Example: PtIr

The second example treats the response of the PtIr cantilever to a force acting in the $\hat{z}$ direction. Figure 2.5 shows the slope of the cantilever, $\theta_z(x)$, as a function of the distance x from its base for a tip–sample force acting in the $\hat{z}$ direction for $F_z = 10$ (solid line), 20 (dashed line), and 30 nN (dotted line). The angular spring constant of the cantilever, k_{θ_z}, is depicted in the figure. Note

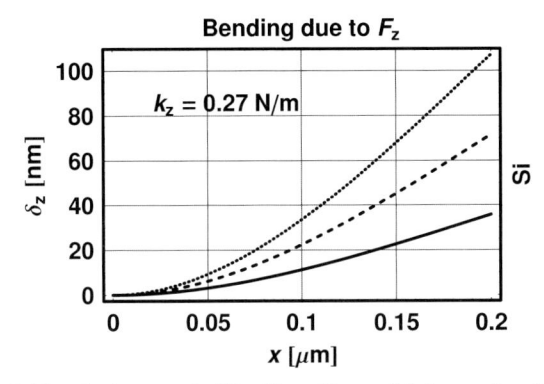

Fig. 2.3 The displacement of the Si cantilever, $\delta_z(x)$, as a function of the distance x from its base, for a tip–sample force acting in the $\hat{z}$ direction for $F_z = 10$ (solid line), 20 (dashed line) and 30 nN (dotted line).

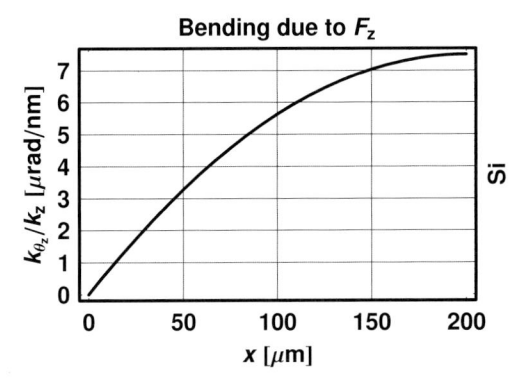

Fig. 2.4 The ratio of the angular and linear spring constants of the Si cantilever, $\zeta_z(x)$, as a function of the distance x from its base, for a tip–sample force acting in the $\hat{z}$ direction.

that, as in the Si cantilever, here too the slope is the steepest next to the base of the cantilever, where it is most likely to break.

Figure 2.6 shows the displacement of the cantilever, $\delta_z(x)$, as a function of the distance x from its base, for a tip–sample force acting in the $\hat{z}$ direction for $F_z = 10$ (solid line), 20 (dashed line), and 30 nN (dotted line). The angular spring constant of the cantilever, k_{θ_z}, is depicted in the figure.

Figure 2.7 shows the ratio of the angular and linear spring constants of the PtIr cantilever, $\zeta_z(x)$, as a function of the distance x from its base.

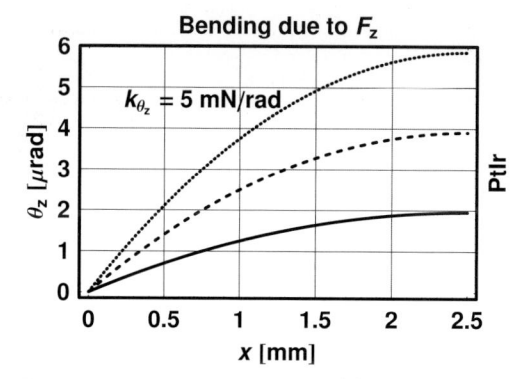

Fig. 2.5 The slope of the PtIr cantilever, $\theta_z(x)$, as a function of the distance x from its base, for a tip–sample force acting in the $\hat{z}$ direction, for $F_z = 10$ (solid line), 20 (dashed line), and 30 nN (dotted line).

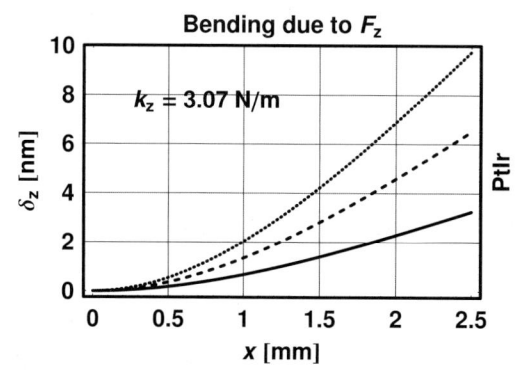

Fig. 2.6 The displacement of the PtIr cantilever, $\delta_z(x)$, as a function of the distance x from its base, for a tip–sample force acting in the $\hat{z}$ direction, for $F_z = 10$ (solid line), 20 (dashed line), and 30 nN (dotted line).

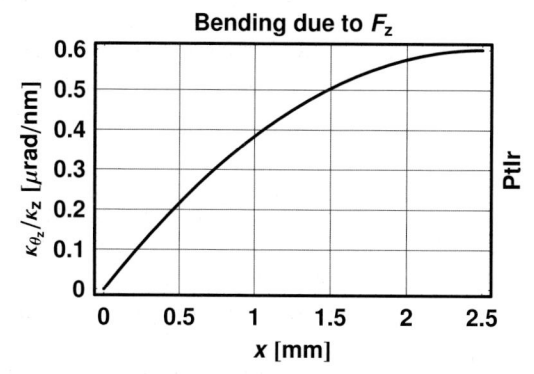

Fig. 2.7 The ratio of the angular and linear spring constants of the PtIr cantilever, $\zeta_z(x)$, as a function of the distance x from its base, for a tip–sample force acting in the $\hat{z}$ direction.

2.3
Buckling Due to F_x

2.3.1
General Equations

Consider now the tip of a cantilever that scans a sample in the $\hat{x}$ direction and ignore the normal tip–sample force, F_z. The scanning will give rise to a tip–sample friction force, F_x, in the $\hat{x}$ direction. The equation of buckling in this case is

$$\frac{\partial^2}{\partial x^2}\delta_x(x) = \frac{M_x}{EI_x}. \tag{2.19}$$

Here, M_x, which is the buckling moment induced by the force F_x, is

$$M_x = h_t F_x, \tag{2.20}$$

where h_t is the tip height. The axial moment of inertia, I_x, is the same as that for F_x,

$$I_x = \frac{1}{12}wd^3. \tag{2.21}$$

Inserting Eqs. (2.20) and (2.21) into Eq. (2.19) yields an explicit, second-order differential equation relating the displacement of the cantilever to the buckling force,

$$\frac{\partial^2}{\partial x^2}\delta_x(x) = 12\frac{1}{Ewd^3}h_t F_x. \tag{2.22}$$

Equation (2.22) will form the basis of the modeling of the response of the cantilever to the force F_x.

2.3.2
Slope

The slope of the cantilever at a point x along its length, $\theta_x(x)$, is obtained by integrating Eq. (2.22),

$$\theta_x(x) = 12\frac{x}{Ewd^3}h_t F_x. \tag{2.23}$$

which, at $x = L$, gives

$$\theta_x(L) = 12\frac{L}{Ewd^3}h_t F_x. \tag{2.24}$$

2.3.3
Angular Spring Constant

The cantilever angular spring constant, $k_{\theta_x}(x)$, is the ratio of the force acting on its tip to the resultant slope of the cantilever at the point x,

$$k_{\theta_x} = \left| \frac{F_x}{\theta_x(x)} \right|. \tag{2.25}$$

Inserting Eq. (2.23) into Eq. (2.25) yields

$$k_{\theta_x} = \frac{1}{12} \frac{Ewd^3}{h_t L}. \tag{2.26}$$

2.3.4
Displacement

The displacement of the cantilever in the $\hat{x}$ direction, $\delta_x(x)$, is obtained by integrating its slope, Eq. (2.23),

$$\delta_x(x) = 6 \frac{x^2}{Ewd^3} h_t F_x. \tag{2.27}$$

2.3.5
Linear Spring Constant

The cantilever linear spring constant, k_x, is defined as the ratio of the force acting on its tip to the resultant displacement at $x = L$,

$$k_x = \left| \frac{F_x}{\delta_x(L)} \right|. \tag{2.28}$$

Inserting Eq. (2.27) into Eq. (2.28) yields

$$k_x = \frac{1}{6} \frac{Ewd^3}{h_t L^2}. \tag{2.29}$$

Note the strong dependence of k_x on the thickness and length of the cantilever. The ratio of the angular and linear spring constants, given by the function $\tilde{\zeta}_x(x)$, is

$$\tilde{\zeta}_x(x) = \frac{k_{\theta_x}(x)}{k_x}. \tag{2.30}$$

Inserting Eq. (2.26) and Eq. (2.29) into Eq. (2.30) gives

$$\tilde{\zeta}_x(x) = \frac{2x}{L^2}. \tag{2.31}$$

2.3.6
Numerical Example: Si

We can now present numerical examples for the response of the Si cantilever to a force acting in the $\hat{x}$ direction. Figure 2.8 shows the slope of the cantilever,

$\theta_x(x)$, as a function of the distance x from its base for $F_x = 10$ (solid line), 20 (dashed line), and 30 nN (dotted line). The angular spring constant of the cantilever, k_{θ_x}, is depicted in the figure.

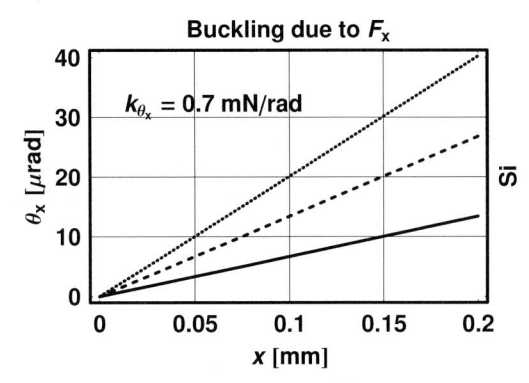

Fig. 2.8 The slope of the Si cantilever, $\theta_x(x)$, as a function of the distance x from its base, for a tip–sample force acting in the $\hat{x}$ direction for $F_x = 10$ (solid line), 20 (dashed line), and 30 nN (dotted line).

Figure 2.9 shows the displacement of the Si cantilever, $\delta_x(x)$, as a function of the distance x from its base for $F_x = 10$ (solid line), 20 (dashed line), and 30 nN (dotted line). Here, the linear spring constant of the cantilever, k_x, is depicted in the figure.

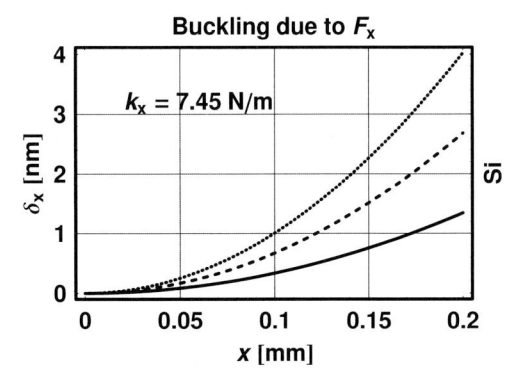

Fig. 2.9 The displacement of the Si cantilever, $\delta_x(x)$ as a function of the distance x from its base, for a tip–sample force acting in the $\hat{x}$ direction for $F_x = 10$ (solid line), 20 (dashed line), and 30 nN (dotted line).

Figure 2.10 shows the ratio of the angular and linear spring constants of the Si cantilever, $\zeta_x(x)$, as a function of the distance x from its base.

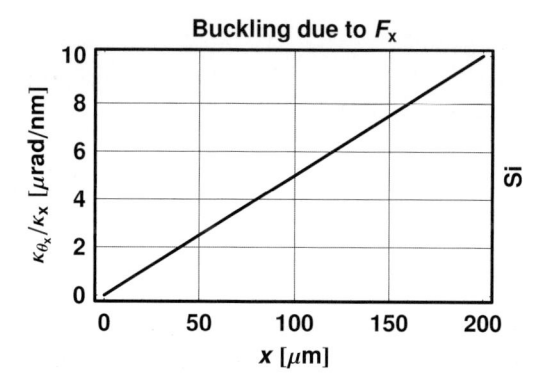

Fig. 2.10 The ratio of the angular and linear spring constants of the Si cantilever, $\zeta_x(x)$, as a function of the distance x from its base, for a tip–sample force acting in the $\hat{x}$ direction.

2.3.7 Numerical Example: PtIr

The second example treats the response of the PtIr cantilever to a force acting in the $\hat{x}$ direction. Figure 2.11 shows the slope of the cantilever, $\theta_x(x)$, as a function of the distance x from its base for $F_x = 10$ (solid line), 20 (dashed line), and 30 nN (dotted line). The angular spring constant of the cantilever, k_{θ_x}, is depicted in the figure.

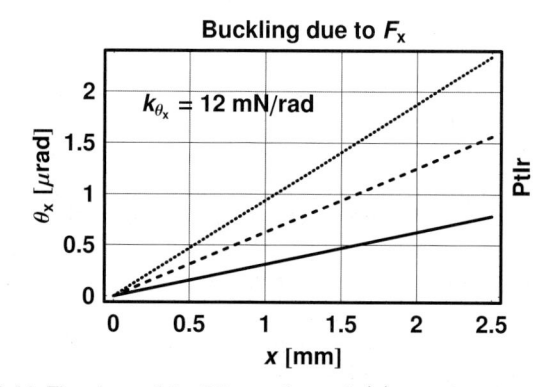

Fig. 2.11 The slope of the PtIr cantilever, $\theta_x(x)$, as a function of the distance x from its base, for a tip–sample force acting in the $\hat{x}$ direction, for $F_x = 10$ (solid line), 20 (dashed line), and 30 nN (dotted line).

Figure 2.12 shows the displacement of the cantilever, $\delta_x(x)$, as a function of the distance x from its base for $F_x = 10$ (solid line), 20 (dashed line), and 30 nN (dotted line). The angular spring constant of the cantilever, k_{θ_x}, is depicted in the figure.

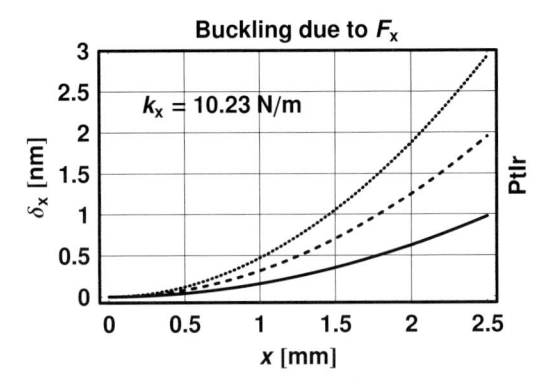

Fig. 2.12 The displacement of the PtIr cantilever, $\delta_x(x)$, as a function of the distance x from its base, for a tip–sample force acting in the $\hat{x}$ direction.

Figure 2.13 shows the ratio of the angular and linear spring constants of the PtIr cantilever, $\zeta_x(x)$, as a function of the distance x from its base.

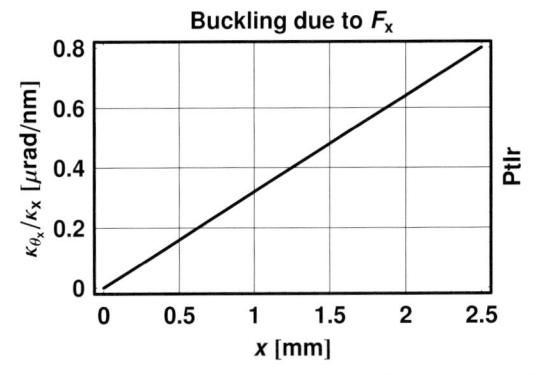

Fig. 2.13 The ratio of the angular and linear spring constants of the PtIr cantilever, $\zeta_x(x)$, as a function of the distance x from its base, for a tip–sample force acting in the $\hat{x}$ direction.

2.4 Twisting Due to F_y

2.4.1 General Equation

The bending moment, T, associated with the neutral axis of the cantilever which is induced by the force F_y, is

$$T = h_t F_y. \tag{2.32}$$

It can be shown that for very thin cantilevers, namely for $d \ll w$, the polar moment of inertia, J, can be approximated by

$$J = \frac{wd^3}{3}. \tag{2.33}$$

To calculate the effect that F_y has on the shape of the cantilever, one also needs to know the shear modulus of its material, G, given by

$$G = \frac{E}{2(1+\nu)}, \tag{2.34}$$

where ν is Poisson's ratio for the cantilever's material.

2.4.2
Slope

It can be shown that the slope of the cantilever at a point x along its length, $\phi_y(x)$, is given by

$$\phi_y(x) = \frac{Tx}{GJ}. \tag{2.35}$$

Inserting Eqs. (2.32) to (2.34) into Eq. (2.35) yields

$$\phi_y(x) = \frac{6h_t x(1+\nu)}{Ewd^3} F_y, \tag{2.36}$$

which, at the free end of the cantilever, where $x = L$, gives

$$\phi_y(L) = \frac{6h_t L(1+\nu)}{Ewd^3} F_y. \tag{2.37}$$

2.4.3
Angular Spring Constant

The cantilever angular spring constant, $k_{\phi_y}(x)$, is defined as the ratio of the force acting on its tip and the resultant slope of the cantilever at the point x,

$$k_{\phi_y} = \left| \frac{F_y}{\phi_y(L)} \right|. \tag{2.38}$$

Inserting Eq. (2.37) into Eq. (2.38) yields

$$k_{\phi_y} = \frac{Ewd^3}{6h_t L(1+\nu)}. \tag{2.39}$$

2.4.4
Numerical Example: Si

We can now present numerical examples for the response of the Si cantilever to a force acting in the $\hat{y}$ direction. Figure 2.14 shows the slope of the cantilever, $\phi_y(x)$, as a function of the distance x from its base for $F_y = 10$ (solid line), 20 (dashed line), and 30 nN (dotted line). The angular spring constant of the cantilever, $k_{\phi_y}(L)$, is depicted in the figure.

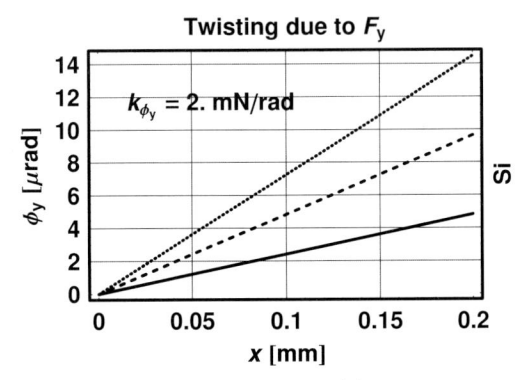

Fig. 2.14 The slope of the Si cantilever, $\phi_y(x)$, as a function of the distance x from its base, for a tip–sample force acting in the $\hat{y}$ direction for $F_y = 10$ (solid line), 20 (dashed line), and 30 nN (dotted line).

2.4.5
Numerical Example: PtIr

The second example treats the response of the PtIr cantilever to a force acting in the $\hat{y}$ direction. Figure 2.15 shows the slope of the cantilever, $\phi_y(x)$, as a function of the distance x from its base for $F_y = 10$ (solid line), 20 (dashed line), and 30 nN (dotted line). The angular spring constant of the cantilever, k_{ϕ_y}, is depicted in the figure.

2.5
Vibrations

2.5.1
Bending Resonance Frequencies

It is important to know the resonance frequency of a given cantilever, since it sets the upper limit to the scanning speed of an atomic force microscope. In the following discussions, we neglect the influence of the mass of the tip and of

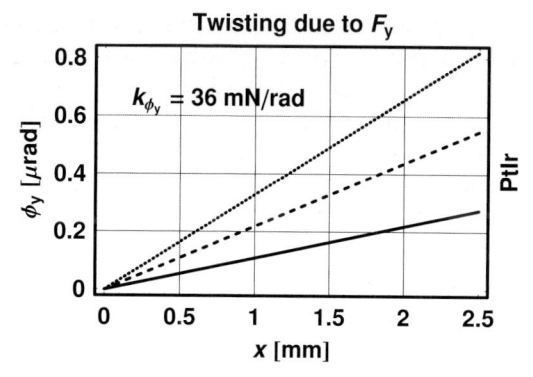

Fig. 2.15 The slope of the PtIr cantilever, $\theta_z(x)$, as a function of the distance x from its base, for a tip–sample force acting in the $\hat{y}$ direction for $F_y = 10$ (solid line), 20 (dashed line), and 30 nN (dotted line).

any cantilever metal coating on the resonance frequency. The time-dependent equation of the displacement of a uniform section cantilever, $\delta_z(x, t)$, neglecting damping effects, is given by the differential equation

$$\rho A \frac{\partial^2}{\partial t^2} \delta_z(x, t) + E I_z \frac{\partial^4}{\partial x^4} \delta_z(x, t) = 0. \tag{2.40}$$

Here, t denotes time, ρ is the mass density of the cantilever material, I_z and E its axial moment of inertia and Young's modulus, respectively, and A its cross-section area. For harmonic motion, Eq. (2.40) reduces to

$$-\omega^2 \rho A \delta_z(x, t) + E I_z \frac{\partial^4}{\partial x^4} \delta_z(x, t) = 0, \tag{2.41}$$

where ω is the driving radial frequency. The solution to this equation, subject to the boundary conditions of a vibrating cantilever, yields a set of resonance frequencies determined by the secular equation

$$\cos \xi_i \cosh \xi_i + 1 = 0, \tag{2.42}$$

where i is the order of the harmonics, $\xi_i = \kappa_i L$, and

$$\kappa_i^4 = \frac{\omega_i^2 \rho}{E I_z}. \tag{2.43}$$

The resonance frequencies, f_i, are readily given by

$$f_i = \frac{1}{2\pi} \left(\frac{\xi_i}{L}\right)^2 \sqrt{\frac{E I_z}{\rho A}}, \tag{2.44}$$

which, for our case, yields

$$f_i = \frac{1}{4\pi}\left(\frac{\zeta_i}{L}\right)^2 d\sqrt{\frac{E}{3\rho}}. \qquad (2.45)$$

Note that for a body with a mass m, suspended by a spring with a spring constant k, the resonance frequency is given by

$$f = \frac{1}{2\pi}\sqrt{\frac{k}{m}}. \qquad (2.46)$$

However, the cantilever behaves as a distributed mass, for which the resonance frequency, f_1, can be approximated by

$$f_1 \approx \frac{1}{2\pi}\sqrt{\frac{k_z}{0.24\rho wdL}}. \qquad (2.47)$$

The approximate resonance frequency, Eq. (2.47), and the first three exact resonance frequencies, Eq. (2.45), for the Si and PtIr cantilevers, are given in Table 2.3. It is important to note that the higher harmonic frequencies are not multiples of the first order one.

Table 2.3 The first three resonance frequencies for the Si and PtIr cantilevers.

		f_i(kHz)	
i	ζ_i	PtIr	Si
1	Approx.	1.731 86	35.593 7
1	1.875 1	1.722 3	35.397 2
2	4.694 09	10.793 5	221.83
3	7.854 76	30.222	621.132

2.5.2
Characteristic Functions

The shape of the vibrating cantilever, described by a characteristic function associated with a particular resonance frequency, is treated in detail in Ref. [4]. The coefficients of the polynomials that represent these characteristic functions are given by a set of parameters, $c_{1,i}$ and $c_{2,i}$, given by

$$c_{1,i} = \frac{\sin(\zeta_i) - \sinh(\zeta_i)}{\cos(\zeta_i) + \cosh(\zeta_i)} \qquad (2.48)$$

and

$$c_{2,i} = \frac{\zeta_i}{L}. \tag{2.49}$$

The resulting shape of the amplitude of the vibrating cantilever, $\delta_z(x,t)$, is given by

$$\delta_{z,i}(x) = \cos(c_{2,i}x) - \cosh(c_{2,i}x) + c_{1,i}[\sin(c_{2,i}x) - \sinh(c_{2,i}x)]. \tag{2.50}$$

The time-dependent shape of the cantilever is the product of the characteristic functions and $\sin(\omega t)$. The characteristic functions for the first three harmonics are shown in Figures 2.16 and 2.17 for the Si and PtIr cantilevers, respectively. The solid, dashed, and dotted lines in these figures refer to cantilever oscillations at f_1, f_2, and f_3, respectively. Note that a cantilever can obtain a shape which is a linear combination of different characteristic functions, depending on the initial conditions and the driving force. For all practical purposes, however, for as long as the cantilever is far away from a surface and its amplitude of vibration is much smaller than its length, its motion can be considered to be linear in distance and driving force.

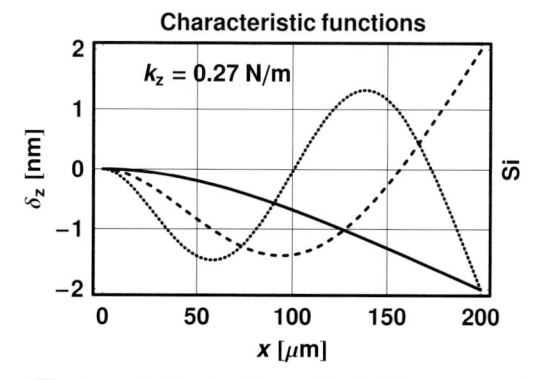

Fig. 2.16 The characteristic functions of the first three harmonics for the Si cantilever.

2.6 Summary of Results

A summary of the equations for the deflection, linear spring constant, slope, and angular spring constants of uniform, rectangular-section cantilevers is given in Table 2.4.

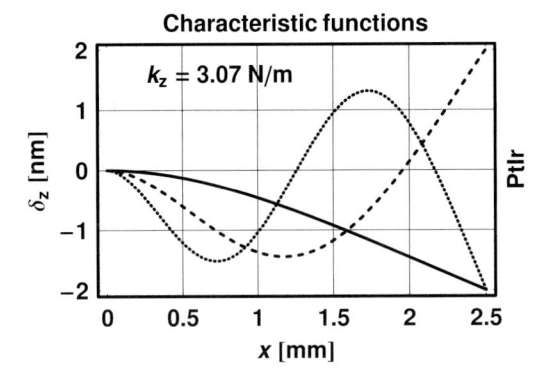

Fig. 2.17 The characteristic functions of the first three harmonics for the PtIr cantilever.

Table 2.4 A summary of the equations for the deflection, linear spring constant, slope, and angular spring constants of uniform, rectangular-section cantilevers.

Bending	Buckling	Twisting
$\delta_z(x) = \dfrac{2(3L-x)x^2 F_z}{Ed^3 w}$	$\delta_x(x) = \dfrac{6x^2 F_x h_t}{Ed^3 w}$	$\phi_y(x) = \dfrac{6x(1-\nu)F_y h_t}{Ed^3 w}$
$k_z = \dfrac{Ed^3 w}{4L^3}$	$k_x = \dfrac{Ed^3 w}{6L^2 h_t}$	$k_{\phi_y} = \dfrac{Ed^3 w}{6L(1-\nu)h_t}$
$\theta_z(x) = \dfrac{6(2L-x)xF_z}{Ed^3 w}$	$\theta_x(x) = \dfrac{12xF_x h_t}{Ed^3 w}$	
$k_{\theta_z} = \dfrac{Ed^3 w}{6L^2}$	$k_{\theta_x} = \dfrac{Ed^3 w}{12Lh_t}$	

▶ **Exercises for Chapter 2**

1. Why is it important to know the resonance frequency of a cantilever when it operates in the contact mode, namely, when it is not vibrated?
2. For which applications is it important to know the linear spring constant and for which is the angular spring constant important?
3. Calculate the spring constants and resonance frequencies for commercially available cantilevers.

References

1 W. C. Young, *Roark's Formulas for Stress and Strain*, McGraw-Hill, New York, 1989.
2 Y. C. Ku, *Deflection of Beams for Spans and Cross Sections*, McGraw-Hill, New York, 1986.
3 P. M. Morse, *Vibration and Sound*, McGraw-Hill, New York 1948.
4 G. Y. Chen, R. J. Warmack, T. Thundat, and D. P. Allison, "Resonance response of scanning force microscopy cantilevers," Rev. Sci. Instrum. **65**, 2532 (1994).

3
Cantilever Conversion Tables

■ Highlights

1. Cantilevers having circular, square, and rectangular sections
2. Conversion tables among the various parameters of each type of cantilever

Abstract

Work in scanning probe microscopy may require the use of uniform cantilevers made of silicon or silicon nitride having rectangular sections, or metallic cantilevers made of wires having circular or square sections. For each type of cantilever, one can write down two coupled equations in terms of the cantilever parameters: Young's modulus, mass density, length, width, thickness (or radius), spring constant, and resonance frequency. The solution to these equations makes it possible to obtain an expression for each one of these parameters in terms of the others. The chapter presents tables of conversion among the parameters of each type of cantilever together with numerical examples.

3.1
Introduction

An atomic force microscope usually employs commercially available cantilevers made of silicon or silicon nitride whose thickness is constant and whose width is either constant or has a triangular shape. For experiments involving electric conductivity, one may choose to employ cantilevers made of metallic wires having circular or square sections, shown in Figure 3.1. The spring constant, k, and the resonance frequency, f, of these cantilevers have to be tailored to a given application that involves the mechanical property of a sample and the speed of scanning. This chapter provides conversion tables for cantilevers having circular, square, and rectangular sections. Each table presents the value of one parameter characterizing the mechanical properties of the particular cantilever in terms of the other parameters. For cantilevers having a circular section, the parameters belong to the set $\{L, r, k, f\}$, where L and r are the length and radius of the cantilever, respectively (see Figure 3.1). For cantilevers having a square section, the parameters belong to the set $\{L, w, k, f\}$, where L and w are the length and width (thickness) of the cantilever, respectively. The accompanied code calculates the value of each of the parameters belonging to each set in terms of two other parameters. For cantilevers having a rectangular section, the parameters belong to the set $\{L, w, d, k, f\}$, where L, w, and d are the length, width, and thickness of the cantilever, respectively. Numerical values for typical parameter sets and for Young's modulus, E, and the mass density, ρ, are also presented, making the conversion tables a useful tool for optimizing the operation of atomic force microscopes.

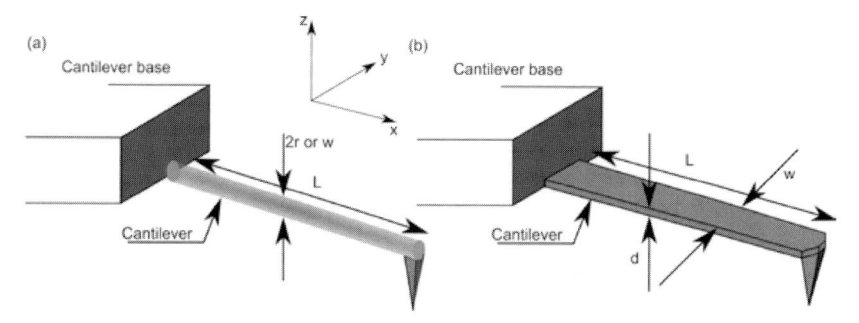

Fig. 3.1 Schematic diagrams of uniform cantilevers having (a) circular and (b) rectangular sections.

3.2
Circular Cantilever

The spring constant of a cantilever made of a metallic wire with a circular section is

$$k = \frac{3\pi}{4} E \frac{r^4}{L^3}, \tag{3.1}$$

and its resonance frequency, in terms of k, is

$$f = \frac{1}{\pi} \sqrt{\frac{k}{\pi \rho r^2 L}}. \tag{3.2}$$

Note that we have used the resonance frequency of a distributed load and approximated it by using $1/4$ rather than 0.24 in the square root. Replacing k in Eq. (3.2) with its value given by Eq. (3.1), yields the value of f in terms of L and r,

$$f = \frac{1}{2\pi} \sqrt{\frac{3E}{\rho} \frac{r}{L^2}}. \tag{3.3}$$

Next, define two material-related constants, c_1 and c_2, by

$$c_1 = \frac{1}{2\pi}\sqrt{\frac{3E}{\rho}} \qquad (3.4)$$

and

$$c_2 = \frac{3\pi}{4}E. \qquad (3.5)$$

Selecting a given cantilever material dictates the values of c_1 and c_2. The resonance frequency and spring constant can therefore be written in terms of c_1 and c_2 as

$$f = c_1\frac{r}{L^2} \qquad (3.6)$$

and

$$k = c_2\frac{r^4}{L^3}, \qquad (3.7)$$

respectively, by the two equations

$$kL^3 = c_2 r^4 \qquad (3.8)$$

and

$$fL^2 = c_1 r, \qquad (3.9)$$

which involve the parameter set $\{L, r, k, f\}$. The solution of these two equations gives the value of each parameter in terms of two others. The six permutations for all the unique pairs of these parameters yield Table 3.1, where each parameter is expressed in terms of two other parameters.

As an example, one can choose the four parameters given in Table 3.2 for a cantilever made of PtIr. The code accompanying this chapter, which generated Table 3.1, calculates two out of the four parameters given by the list $\{L, r, k, f\}$, once the other two parameters are given by a replacement rule.

3.3
Square Cantilever

The spring constant of a cantilever made of a metallic wire having a square section is

$$k = \frac{1}{4}E\frac{w^4}{L^3}, \qquad (3.10)$$

3.3 Square Cantilever

Table 3.1 The permutations for all the unique pairs of parameters belonging to the set $\{L, r, k, f\}$, where each parameter is expressed in terms of the others. As an example, we choose the four parameters given in Table 3.2, and a cantilever made of PtIr.

$$f(L,r) = \frac{rc_1}{L^2} \qquad k(L,r) = \frac{r^4 c_2}{L^3}$$

$$f(k,r) = \frac{c_1 \left(\frac{k}{c_2}\right)^{2/3}}{r^{5/3}} \qquad L(k,r) = r^{4/3} \left(\frac{c_2}{k}\right)^{1/3}$$

$$f(L,k) = \frac{c_1 \left(\frac{k}{c_2}\right)^{1/4}}{L^{5/4}} \qquad r(L,k) = L^{3/4} \left(\frac{k}{c_2}\right)^{1/4}$$

$$L(f,k) = \left(\frac{c_1}{f}\right)^{4/5} \left(\frac{k}{c_2}\right)^{1/5} \qquad r(f,k) = \left(\frac{c_1}{f}\right)^{3/5} \left(\frac{k}{c_2}\right)^{2/5}$$

$$k(f,r) = r^{5/2} \left(\frac{f}{c_1}\right)^{3/2} c_2 \qquad L(f,r) = \sqrt{\frac{rc_1}{f}}$$

$$k(f,L) = \frac{f^4 L^5 c_2}{c_1^4} \qquad r(f,L) = \frac{fL^2}{c_1}$$

Table 3.2 The parameters of the cantilever made of a PtIr wire having a circular section.

$$E = 240 \, \text{GN/m}^2$$
$$\rho = 21\,620 \, \text{kg/m}^3$$
$$L = 644.933 \, \mu\text{m}$$
$$r = 4.809\,39 \, \mu\text{m}$$
$$k = 1 \, \text{N/m}$$
$$f = 10 \, \text{kHz}$$

and its resonance frequency is

$$f = \frac{1}{\pi} \sqrt{\frac{k}{\rho w^2 L}} . \tag{3.11}$$

Replacing k in Eq. (3.11) with its value given by Eq. (3.10), yields the value of f in terms of L and w,

$$f = \frac{1}{2\pi} \sqrt{\frac{E}{\rho} \frac{w}{L^2}} . \tag{3.12}$$

Next, define two material constants, c_1 and c_2, by

$$c_1 = \frac{1}{2\pi} \sqrt{\frac{E}{\rho}} \tag{3.13}$$

and

$$c_2 = \frac{E}{4}. \tag{3.14}$$

Selecting a given cantilever material dictates the values of c_1 and c_2. The resonance frequency and spring constant can therefore be written in terms of c_1 and c_2 as

$$f = c_1 \frac{w}{L^2} \tag{3.15}$$

and

$$k = c_2 \frac{w^4}{L^3}, \tag{3.16}$$

respectively. We can now write the two equations

$$kL^3 = c_2 w^4 \tag{3.17}$$

and

$$fL^2 = c_1 w, \tag{3.18}$$

which involve the parameter set $\{L, w, k, f\}$. The solution of these two equations gives the value of each parameter in terms of the other two. The six permutations for all the unique pairs of these parameters yield Table 3.3, where each parameter is expressed in terms of the other two parameters.

As an example, one can choose the four parameters given in Table 3.4 for a cantilever made of PtIr. The code accompanying this chapter, which generated Table 3.3, calculates two out of the four parameters given by the list $\{L, w, k, f\}$, once the other two parameters are given by a replacement rule.

3.4
Rectangular Cantilever

The spring constant of a cantilever having a rectangular section is

$$k = \frac{1}{4} E \frac{wd^3}{L^3}, \tag{3.19}$$

and its resonance frequency is

$$f = \frac{1}{\pi} \sqrt{\frac{k}{\rho w d L}}. \tag{3.20}$$

3.4 Rectangular Cantilever

Table 3.3 The permutations for all the unique pairs belonging to the set $\{L, w, k, f\}$, where each parameter is expressed in terms of the other three parameters.

$$f(L,w) = \frac{wc_1}{L^2} \qquad k(L,w) = \frac{w^4 c_2}{L^3}$$

$$f(k,w) = \frac{c_1 \left(\frac{k}{c_2}\right)^{2/3}}{w^{5/3}} \qquad L(k,w) = w^{4/3}\left(\frac{c_2}{k}\right)^{1/3}$$

$$f(L,k) = \frac{c_1 \left(\frac{k}{c_2}\right)^{1/4}}{L^{5/4}} \qquad w(L,k) = L^{3/4}\left(\frac{k}{c_2}\right)^{1/4}$$

$$L(f,k) = \left(\frac{c_1}{f}\right)^{4/5}\left(\frac{k}{c_2}\right)^{1/5} \qquad w(f,k) = \left(\frac{c_1}{f}\right)^{3/5}\left(\frac{k}{c_2}\right)^{2/5}$$

$$k(f,w) = w^{5/2}\left(\frac{f}{c_1}\right)^{3/2} c_2 \qquad L(f,w) = \sqrt{\frac{wc_1}{f}}$$

$$k(f,L) = \frac{f^4 L^5 c_2}{c_1^4} \qquad w(f,L) = \frac{fL^2}{c_1}$$

Table 3.4 The parameters of the cantilever made of a PtIr wire having a square section.

$E = 240\,\text{GN/m}^2$
$\rho = 21\,620\,\text{kg/m}^3$
$L = 650.908\,\mu\text{m}$
$w = 8.485\,18\,\mu\text{m}$
$k = 1\,\text{N/m}$
$f = 10\,\text{kHz}$

Replacing k in Eq. (3.20) with its value given by Eq. (3.19), yields the value of f in terms of L and d,

$$f = \frac{1}{2\pi}\sqrt{\frac{E}{\rho}\frac{d}{L^2}}. \qquad (3.21)$$

Next, using the constants c_1 and c_2, Eqs (3.13) and (3.14), to describe the resonance frequency and spring constant, yields

$$f = c_1 \frac{d}{L^2} \qquad (3.22)$$

and

$$k = c_2 \frac{wd^3}{L^3}. \qquad (3.23)$$

We can now write the two equations

$$kL^3 = c_2 w d^3 \tag{3.24}$$

and

$$fL^2 = c_1 d, \tag{3.25}$$

which involve the parameter set $\{L, d, w, k, f\}$. The solution of these two equations gives the value of each parameter in terms of the others. The six permutations for all the unique pairs of these parameters yield Table 3.3, where each parameter is expressed in terms of the other two. The permutations for all the unique pairs of these parameters yield Table 3.5, where each parameter is expressed in terms of the other three parameters.

Table 3.5 The permutations for all the unique pairs belonging to the set $\{L, d, w, k, f\}$, where each parameter is expressed in terms of the others.

$f(d, w, L) = \dfrac{dc_1}{L^2}$	$k(d, w, L) = \dfrac{d^3 w c_2}{L^3}$
$f(k, d, w) = \dfrac{c_1 \left(\frac{k}{w c_2}\right)^{2/3}}{d}$	$L(k, d, w) = d \left(\dfrac{w c_2}{k}\right)^{1/3}$
$f(k, L, d) = \dfrac{dc_1}{L^2}$	$w(k, L, d) = \dfrac{kL^3}{d^3 c_2}$
$L(k, d, f) = \sqrt{\dfrac{dc_1}{f}}$	$w(k, d, f) = \dfrac{k \left(\frac{c_1}{df}\right)^{3/2}}{c_2}$
$k(d, f, w) = w \left(\dfrac{df}{c_1}\right)^{3/2} c_2$	$L(d, f, w) = \sqrt{\dfrac{dc_1}{f}}$
$k(f, L, w) = \dfrac{f^3 L^3 w c_2}{c_1^3}$	$d(f, L, w) = \dfrac{fL^2}{c_1}$
$w(k, f, L) = \dfrac{k c_1^3}{f^3 L^3 c_2}$	$d(k, f, L) = \dfrac{fL^2}{c_1}$
$L(k, f, w) = \dfrac{c_1 \left(\frac{k}{w c_2}\right)^{1/3}}{f}$	$d(k, f, w) = \dfrac{c_1 \left(\frac{k}{w c_2}\right)^{2/3}}{f}$

As an example, one can choose the five parameters given in Table 3.6 for a cantilever made of silicon.

The correctness of the equations given in Table 3.5 is verified by using the parameters given in Table 3.6, as shown in Table 3.7.

▶ **Exercises for Chapter 3**

1. Make a conversion table for a cantilever having a triangular section.

Table 3.6 The parameters of the silicon cantilever having a rectangular section.

$E = 179\,\text{GN}/\text{m}^2$
$\rho = 2330\,\text{kg}/\text{m}^3$
$L = 200\,\mu\text{m}$
$w = 50\,\mu\text{m}$
$d = 1\,\mu\text{m}$
$k = 0.279\,687\,\text{N}/\text{m}$
$f = 34.8746\,\text{kHz}$

Table 3.7 Verification of the correctness of the equations given in Table 3.5 by using the parameters given in Table 3.6

$f(d, w, L) = 34\,874.6$	$k(d, w, L) = 0.279\,687$
$f(k, d, w) = 34\,874.5$	$L(k, d, w) = 0.000\,2$
$f(k, L, d) = 34\,874.6$	$w(k, L, d) = 0.000\,049\,999\,9$
$L(k, d, f) = 0.000\,2$	$w(k, d, f) = 0.000\,049\,999\,8$
$k(d, f, w) = 0.279\,688$	$L(d, f, w) = 0.000\,2$
$k(f, L, w) = 0.279\,689$	$d(f, L, w) = 1 \times 10^{-6}$
$w(k, f, L) = 0.000\,049\,999\,7$	$d(k, f, L) = 1 \times 10^{-6}$
$L(k, f, w) = 0.000\,2$	$d(k, f, w) = 9.999\,98 \times 10^{-7}$

2. Add to the conversion tables a parameter representing the point mass of a tip at the end of the cantilever.

References

1. W. C. Young, *Roark's Formulas for Stress and Strain*, McGraw-Hill, New York, 1989.
2. D. Sarid, *Scanning Force Microscopy with Applications to Electric, Magnetic and Atomic Forces*, Oxford University Press, New York, 1994.

4
V-Shaped Cantilevers

■ Highlights

1. Effects of bending and buckling
2. Displacement and slope
3. Linear and angular spring constants
4. Resonance frequencies and characteristic functions

Abstract

We dealt previously with the mechanical properties of uniform cantilevers having a rectangular section. The second types of cantilevers, also extensively used in atomic force microscopy, are V-shaped with a uniform thickness. The bending and buckling of these V-shaped cantilevers, their resonance frequencies, and their characteristic functions are presented in this chapter both algebraically and numerically.

4.1
Introduction

Figure 4.1 shows a typical solid, V-shaped cantilever having a constant thickness that is extensively used in atomic force microscopy. The purpose of using V-shaped cantilevers, rather than uniform, rectangular-section ones, is to minimize buckling and twisting effects due to tip–sample friction forces. The geometry of the V-shaped cantilever, shown in Figure 4.1, has its length along the $\hat{x}$ direction with its tip pointing in the $-\hat{z}$ direction, and the sample surface positioned in the $\hat{x}$–$\hat{y}$ plane. The cantilever thickness, width, and length are denoted by d, B, and L, respectively, L_1 is the length of its recess, and w is the width of each of its two arms. Note that for $L_1 = 0$, one obtains a triangular cantilever having no recess, while for $L_1 < L$, the recess can be either a triangle or a trapezoid whose short side is denoted by w_1. The height of the tip and the distance of its center from the free end of the cantilever are denoted by

h_t and δ, respectively. The operation of the V-shaped cantilever and the definitions and notations of its slope, displacement, and angular and linear spring constants are similar to those whose section is rectangular, as described in the previous chapter. In the following sections we treat the bending and buckling of the cantilever due to the tip–sample forces F_y and F_z that act in the $\hat{y}$ and $\hat{z}$ directions, respectively. Unlike uniform cantilevers, here it is important to account for the position of the tip, δ, because the free end of the cantilever is a sharp point. The numerical results presented in this chapter will be associated with $F_i = 10$ (solid line), 20 (dashed line), and 30 nN (dotted line), for $i = y$ and z. The resonance frequencies of the cantilever with their associated characteristic functions are presented for the case where the AFM operates in the noncontact mode.

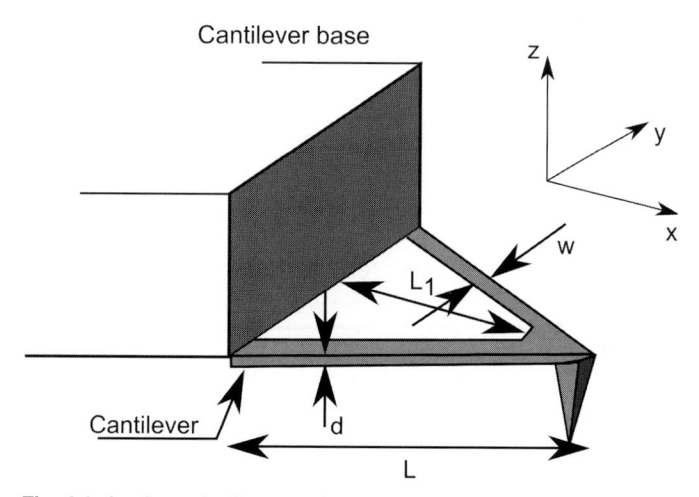

Fig. 4.1 A schematic diagram of a uniform V-shaped cantilever.

A V-shaped cantilever can be viewed as a combination of a triangle attached by two uniform arms to a base. Such a description enables one to arrive at an approximate solution to the response of the cantilever to the tip–sample forces. This solution is obtained by combining the results for a triangular-shaped cantilever with those for two uniform ones. The chapter is therefore divided into two parts: the first part deals with a triangular-shaped cantilever, denoted by the subscript "T," and the second one deals with the V-shaped cantilever, denoted by "V." The theory presented in this chapter is accompanied by numerical examples that relate to these two types of cantilevers. The geometry and material properties of the V-shaped cantilever, which also pertain to the triangular-shaped one, are given in Table 4.1.

Table 4.1 The parameters of the V-shaped cantilever.

$L = 202\,\mu m$
$L_1 = 116\,\mu m$
$B = 205\,\mu m$
$w = 40.2\,\mu m$
$d = 0.54\,\mu m$
$h_t = 3.5\,\mu m$
$\delta = 7.1\,\mu m$
$E = 179\,GN/m^2$
$\rho = 2330\,kg/m^3$

4.2
Bending Due to F_z: Triangular Shape

4.2.1
General Equations

The geometry of the cantilever considered in this section and the next one is a triangle that is similar to the one shown in Figure 4.1, except that the recess is eliminated. We consider a tip, positioned at a distance δ from the free end of the cantilever, that scans a sample in an arbitrary direction in the $\hat{x}$–$\hat{y}$ plane. Let a bending force, F_z, act on the tip in a direction perpendicular to the cantilever body in the $\hat{z}$ direction. Here, $\delta_{z_T}(x)$ denotes the displacement of the cantilever in the $\hat{z}$ direction at a point x along its body, where the subscript denotes the direction of this force. The force F_z in this case arises from changes in the local topography of the sample, which, during the scanning process, raises or lowers the tip, thus either relaxing or stressing the cantilever. The bending moment at a point x along the length of the cantilever, $M_{z_T}(x)$, due to the force, F_z, acting on the tip, is

$$M_{z_T}(x) = (L - x)F_z. \tag{4.1}$$

In contrast to the case of a uniform cantilever, the axial moment of inertia for a triangular cantilever, $I_{z_T}(x)$, is not constant but is rather a function of x,

$$I_{z_T}(x) = (L - x)\frac{Bd^3}{12L}. \tag{4.2}$$

The equation of bending for this case is

$$\frac{\partial^2}{\partial x^2}\delta_{z_T}(x) = \frac{M_{z_T}(x)}{EI_{z_T}(x)}. \tag{4.3}$$

Inserting Eq. (4.1) and Eq. (4.2) into Eq. (4.3) yields an explicit, second-order differential equation that relates the displacement of the cantilever to the bending force,

$$\frac{\partial^2}{\partial x^2}\delta_{z_T}(x) = \frac{12L}{EBd^3}F_z. \qquad (4.4)$$

Equation (4.4) will form the basis of the modeling of the response of the triangular cantilever to F_z.

4.2.2
Slope

The slope of the cantilever, $\theta_{z_T}(x)$, is obtained by integrating Eq. (4.4),

$$\theta_{z_T}(x) = \frac{12Lx}{EBd^3}F_z. \qquad (4.5)$$

Note that the slope is linear in x because Eq. (4.1) and Eq. (4.2) have a similar functional form. At $x = L - \delta$, one obtains

$$\theta_{z_T}(L - \delta) = 12\frac{L(L-\delta)}{EBd^3}F_z. \qquad (4.6)$$

4.2.3
Angular Spring Constant

We define an angular spring constant, $k_{\theta_{z_T}}(x)$, as the ratio of the force acting on the tip to the resultant slope of the cantilever at the point x,

$$k_{\theta_{z_T}}(x) = \left|\frac{F_z}{\theta_{z_T}(x)}\right|. \qquad (4.7)$$

Inserting Eq. (4.5) into Eq. (4.7) yields

$$k_{\theta_{z_T}}(x) = \frac{EBd^3}{12Lx}. \qquad (4.8)$$

For $x = L - \delta$, one obtains a particular value of the angular spring constant, $k_{\theta_{z_T}}$,

$$k_{\theta_{z_T}} = \frac{EBd^3}{12L(L-\delta)}. \qquad (4.9)$$

4.2.4
Displacement

The displacement of the cantilever in the $\hat{z}$ direction, $\delta_{z_T}(x)$, is obtained by integrating its slope, Eq. (4.5),

$$\delta_{z_T}(x) = \frac{6Lx^2}{EBd^3}F_z. \qquad (4.10)$$

At $x = L - \delta$, the displacement is

$$\delta_{z_T}(L - \delta) = \frac{6L(L-\delta)^2}{EBd^3} F_z. \tag{4.11}$$

4.2.5
Linear Spring Constant

The cantilever linear spring constant, k_{z_T}, is defined as the ratio of the force acting on its tip to the resultant displacement at $x = L - \delta$,

$$k_{z_T} = \left| \frac{F_z}{\delta_{z_T}(L-\delta)} \right|. \tag{4.12}$$

Inserting Eq. (4.11) into Eq. (4.12) yields

$$k_{z_T} = \frac{Bd^3 E}{6L(L-\delta)^2}. \tag{4.13}$$

Note the strong dependence of k_{z_T} on the thickness and length of the cantilever and on the value of δ. Remember that tip–sample forces always act at a distance δ away from the cantilever's free end, while the beam of the AFM laser diode is incident at another point, x. It is therefore of interest to examine the ratio of the angular and linear spring constants, $\xi_{z_T}(x)$,

$$\xi_{z_T}(x) = \frac{k_{\theta_{z_T}}(x)}{k_{z_T}}. \tag{4.14}$$

Inserting Eq. (4.9) and Eq. (4.13) into Eq. (4.14) gives

$$\xi_z(x) = \frac{(L-\delta)^2}{2x}. \tag{4.15}$$

Equation (4.15) can be used to calibrate the operation of the AFM in terms of the position x at which the laser beam is incident on the cantilever.

4.2.6
Numerical Examples

We now present numerical examples for the response of the cantilever to a force acting in the $\hat{z}$ direction. Figure 4.2 shows the slope of the cantilever, $\theta_{z_T}(x)$, as a function of the distance x from its base for $F_z = 10$ (solid line), 20 (dashed line), and 30 nN (dotted line). The angular spring constant of the cantilever, $k_{\theta_{z_T}}$, is depicted in the figure. Note that the slope is constant along the length of the cantilever.

4.2 Bending Due to F_n: Triangular Shape

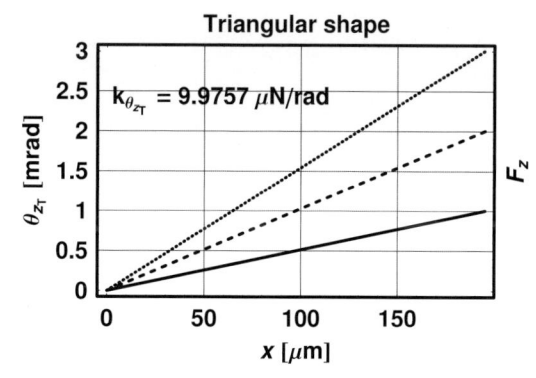

Fig. 4.2 The slope of the triangular cantilever, $\theta_{z_T}(x)$, as a function of the distance x from its base for $F_z = 10$ (solid line), 20 (dashed line), and 30 nN (dotted line). The angular spring constant, $k_{\theta_{z_T}}$, is depicted in the figure.

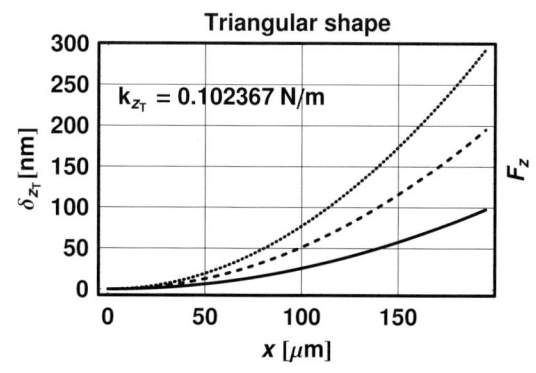

Fig. 4.3 The displacement of the triangular cantilever, $\delta_{z_T}(x)$, as a function of the distance x from its base for $F_z = 10$ (solid line), 20 (dashed line), and 30 nN (dotted line). The linear spring constant, k_{z_T}, is depicted in the figure.

Figure 4.3 shows the displacement of the cantilever, $\delta_{z_T}(x)$, as a function of the distance x from its base. The angular spring constant of the cantilever, $k_{\theta_{z_T}}$, is depicted in the figure.

Figure 4.4 shows the ratio of the linear and angular spring constants of the triangular cantilever, $\zeta_{z_T}(x)$, as a function of the distance x from its base. This function is used by the software of the AFM to calibrate the tip–sample force using the specifications of the particular cantilever in use.

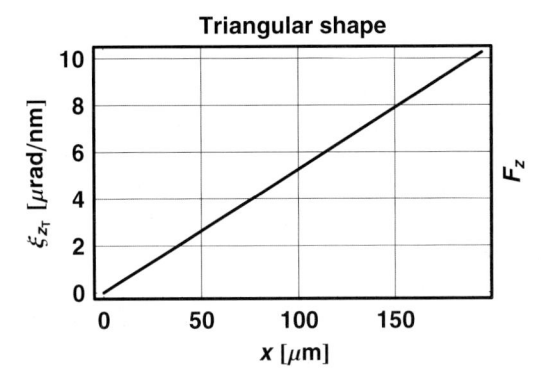

Fig. 4.4 The ratio of the linear and angular spring constants of the triangular cantilever, $\xi_{z_T}(x)$, as a function of the distance x from its base.

4.3
Buckling due to F_x: Triangular Shape

4.3.1
General Equations

Consider now the case where the tip of the cantilever scans a sample in the $\hat{x}$ direction and ignore the normal tip–sample force, F_z. The scanning will give rise to a tip–sample friction force, F_x, acting in the $\hat{x}$ direction. The resultant buckling moment, M_{x_T}, is

$$M_{x_T} = h_t F_x . \tag{4.16}$$

The axial moment of inertia, $I_z(x)$, is the same as that for the bending case,

$$I_{x_T}(x) = (L - x)\frac{Bd^3}{12L} . \tag{4.17}$$

The equation of buckling in this case is

$$\frac{\partial^2}{\partial x^2}\delta_{x_T}(x) = \frac{M_{x_T}}{EI_{x_T}(x)} . \tag{4.18}$$

Inserting Eq. (4.16) and Eq. (4.17) into Eq. (4.18) yields an explicit, second-order differential equation relating the displacement of the cantilever to the buckling force,

$$\frac{\partial^2}{\partial x^2}\delta_{x_T}(x) = \frac{12L}{EBd^3(L-x)}h_t F_x . \tag{4.19}$$

Equation (4.19) will form the basis of the modeling of the response of the triangular cantilever to F_x.

4.3.2
Slope

The slope of the cantilever at a point x along its length, $\theta_{x_T}(x)$, is obtained by integrating Eq. (4.19),

$$\theta_{x_T}(x) = \frac{12L}{EBd^3} \log\left[\frac{L}{L-x}\right] h_t F_x, \qquad (4.20)$$

which diverges at $x = L$. For $x = L - \delta$, however, one obtains

$$\theta_{x_T}(L-\delta) = \frac{12L}{EBd^3} \log\left[\frac{L}{\delta}\right] h_t F_x, \qquad (4.21)$$

4.3.3
Angular Spring Constant

The cantilever angular spring constant, $k_{\theta_{x_T}}(x)$, is the ratio of the force acting on its tip to the resultant slope of the cantilever at the point x,

$$k_{\theta_{x_T}} = \left|\frac{F_x}{\theta_{x_T}(L-\delta)}\right|. \qquad (4.22)$$

Inserting Eq. (4.20) into Eq. (4.22) yields

$$k_{\theta_{x_T}}(x) = \frac{EBd^3}{12 h_t L \log\left(\frac{L}{L-x}\right)}. \qquad (4.23)$$

For $x = L - \delta$, one obtains the particular value of the angular spring constant, $k_{\theta_{x_T}}$,

$$k_{\theta_{x_T}} = \frac{EBd^3}{12 h_t L \log\left(\frac{L}{\delta}\right)}. \qquad (4.24)$$

4.3.4
Displacement

The displacement of the cantilever in the $\hat{x}$ direction, $\delta_{x_T}(x)$, is obtained by integrating its slope, Eq. (4.20), yielding

$$\delta_{x_T}(x) = \frac{12L}{EBd^3}\left[x + (x-L)\log\left(\frac{L}{L-x}\right)\right] h_t F_x. \qquad (4.25)$$

At $x = L - \delta$, the displacement is

$$\delta_{x_T}(L-\delta) = \frac{12L}{EBd^3}\left[L - \delta - \delta \log\left(\frac{L}{\delta}\right)\right] h_t F_x. \qquad (4.26)$$

4.3.5
Linear Spring Constant

The cantilever linear spring constant, k_{x_T}, is defined as the ratio of the force acting on its tip to the resultant displacement at $x = L - \delta$,

$$k_{x_T} = \left| \frac{F_x}{\delta_{x_T}(L-\delta)} \right|. \tag{4.27}$$

Inserting Eq. (4.25) into Eq. (4.27) yields

$$k_{x_T} = \frac{EBd^3}{12h_t L \left[L - \delta - \delta \log\left(\frac{L}{\delta}\right)\right]}. \tag{4.28}$$

Note the strong dependence of k_{x_T} on the thickness and length of the cantilever and on the value of δ. The ratio of the angular and linear spring constants, $\tilde{\zeta}_{x_T}(x)$, is

$$\tilde{\zeta}_{x_T}(x) = \frac{k_{\theta_{x_T}}(x)}{k_{x_T}}. \tag{4.29}$$

Inserting Eqs. (4.24) and (28) into Eq. (4.29) gives

$$\tilde{\zeta}_{x_T}(x) = \frac{\log\left(\frac{L}{L-x}\right)}{L - \delta - \delta \log\left(\frac{L}{\delta}\right)}. \tag{4.30}$$

4.3.6
Numerical Examples

The slope of the triangular cantilever, $\theta_{x_T}(x)$, as a function of the distance x from its base for $F_x = 10$ (solid line), 20 (dashed line), and 30 nN (dotted line) is shown in Figure 4.5. The linear spring constant, k_{x_T}, is depicted in the figure.

Figure 4.6 shows the displacement of the triangular cantilever, $\delta_{x_T}(x)$, as a function of the distance x from its base for $F_x = 10$ (solid line), 20 (dashed line), and 30 nN (dotted line). The value of the angular spring constant, k_{x_T}, is depicted in the figure.

Figure 4.7 shows the ratio of the linear and angular spring constants of the triangular cantilever, $\tilde{\zeta}_{x_T}(x)$, as a function of the distance x from its base. This function is used by the software of the AFM to calibrate the tip–sample force using the specifications of the particular cantilever in use.

4.3 Buckling due to F_x: Triangular Shape

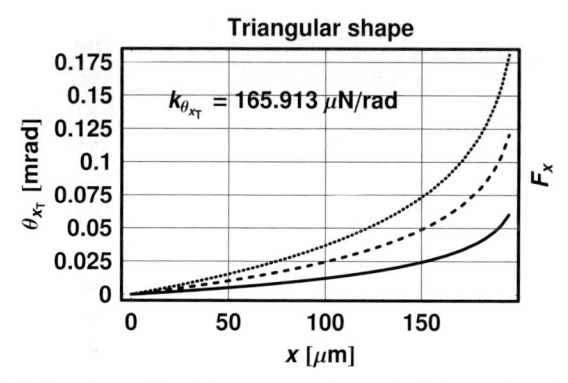

Fig. 4.5 The slope of the triangular cantilever, $\theta_{x_T}(x)$, as a function of the distance x from its base for $F_x = 10$ (solid line), 20 (dashed line), and 30 nN (dotted line). The linear spring constant, k_{x_T}, is depicted in the figure.

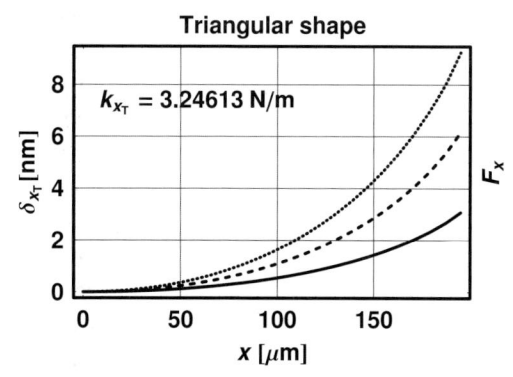

Fig. 4.6 The displacement of the triangular cantilever, $\delta x_T(x)$, as a function of the distance x from its base for $F_x = 10$ (solid line), 20 (dashed line), and 30 nN (dotted line). The linear spring constant, k_{x_T}, is depicted in the figure.

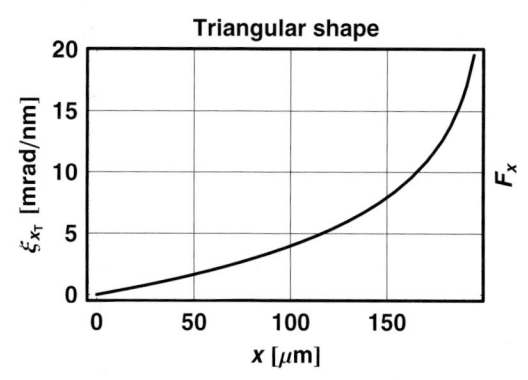

Fig. 4.7 The ratio of the linear and angular spring constants of the triangular cantilever, $\zeta_{x_T}(x)$, as a function of the distance x from its base.

4.4
Bending due to F_z: V Shape

4.4.1
General Equations

The next two sections deal with a V-shape cantilever, shown in Figure 4.1, that has a finite recess. Here, a bending force F_z acts on the tip in a direction perpendicular to the cantilever body. Let $\delta_{zV}(x)$ denote the displacement of the cantilever in the $\hat{z}$ direction at a point x along its body. The bending moment at a point x along the length of the cantilever, $M_{zV}(x)$, is given by

$$M_{zV}(x) = (L-x)F_z. \tag{4.31}$$

In contrast to the case of a triangular cantilever, the solution to the V-shaped one is more difficult, and requires an approximation to obtain a simple analytical expression. The approach will be to decompose the V-shaped cantilever into two components, one in the form of a triangle and the other in the form of two uniform arms. The first component is realized in the range $0 < x < L_1$ and the second one in the range $L_1 < x < L$, denoted by the subscripts 1 and 2, respectively. The axial moment of inertia for the first component is given by considering the two arms as two parallel uniform cantilevers, for which the axial moment of inertia, I_{z_1}, is

$$I_{z_1} = \frac{wd^3}{6}. \tag{4.32}$$

The equation of motion for $0 < x < L_1$ is therefore

$$\frac{\partial^2}{\partial x^2}\delta_{z_1}(x) = \frac{M_{zV}(x)}{EI_{z_1}}, \tag{4.33}$$

which, by using Eqs. (4.31) and (4.32), is given by

$$\frac{\partial^2}{\partial x^2}\delta_{z_1}(x) = 6\frac{L-x}{Ewd^3}F_z. \tag{4.34}$$

The axial moment of inertia for $L_1 < x < L$ is

$$I_{z_2}(x) = \frac{(L-x)Bd^3}{12L}. \tag{4.35}$$

The equation of bending for this range is

$$\frac{\partial^2}{\partial x^2}\delta_{z_2}(x) = \frac{M_{zV}(x)}{EI_{z_2}(x)}. \tag{4.36}$$

Inserting Eqs. (4.31) and (4.35) into Eq. (4.36) yields

$$\frac{\partial^2}{\partial x^2}\delta_{z_2}(x) = \frac{12L}{EBd^3}F_z. \tag{4.37}$$

Equation (4.34) and Eq. (4.37) will form the basis of the modeling of the response of the V-shaped cantilever to F_z.

4.4.2
Slope

The slope of the cantilever, $\theta_{z_V}(x)$, for $0 < x < L_1$, is obtained by integrating Eq. (4.34),

$$\theta_{z_1}(x) = \frac{3(2L-x)x}{Ewd^3} F_z. \tag{4.38}$$

For $L_1 < x < L$, the slope, $\theta_{z_2}(x)$, is obtained by adding the contribution of $\theta_{z_1}(L_1)$ to the integral of Eq. (4.37),

$$\theta_{z_2}(x) = \theta_{z_1}(L_1) + \int_{L_1}^{x} \frac{M_{z_V}(x)}{EI_{z_2}(x)} dx. \tag{4.39}$$

Using Eqs. (4.31), (4.35), and (4.38) in Eq. (4.39) results in

$$\theta_{z_2}(x) = 3\frac{B(2L-L_1)L_1 + 4Lw(x-L_1)}{EwBd^3} F_z. \tag{4.40}$$

Consequently, the slope across the whole cantilever, namely for $0 < x < L$, denoted by $\theta_{z_V}(x)$, takes the form of Eq. (4.38) or Eq. (4.40), depending on their range of applicability.

4.4.3
Angular Spring Constant

The angular spring constant, $k_{\theta_{z_2}}(x)$, is given by the ratio of the force acting on the tip to the resultant slope of the cantilever at the point x,

$$k_{\theta_{z_V}}(x) = \left| \frac{F_z}{\theta_{z_V}(x)} \right|. \tag{4.41}$$

The point x at which the laser diode beam is incident on the cantilever is usually in the range $L_1 < x < L - \delta$. In this case, one obtains for $k_{\theta_{z_V}}(x)$, after inserting Eq. (4.40) into Eq. (4.41),

$$k_{\theta_{z_V}}(x) = \frac{1}{3} \frac{EwBd^3}{B(2L-L_1)L_1 + 4Lw(x-L_1)}. \tag{4.42}$$

4.4.4
Displacement

For the range $0 < x < L_1$, the displacement, $\delta_{z_1}(x)$, is obtained by integrating Eq. (4.38),

$$\delta_{z_1}(x) = \frac{(3L-x)x^2}{Ewd^3} F_z. \tag{4.43}$$

For $L_1 < x < L$, the displacement is obtained by adding the contribution of $\delta_{z_1}(L_1)$ to the integral of Eq. (4.40),

$$\delta_{z_2}(x) = \delta_{z_1}(L_1) + \int_{L_1}^{x} \theta_{z_2}(x) \, dx. \tag{4.44}$$

Using Eqs. (4.40) and (4.43) into Eq. (4.44), one obtains

$$\delta_{z_2}(x) = \frac{6Lw(L_1-x)^2 + BL_1(-3LL_1 + 2L_1^2 + 6Lx - 3L_1 x)}{EwBd^3} F_z. \tag{4.45}$$

4.4.5
Linear Spring Constant

The linear spring constant, k_{z_V}, is defined by

$$k_{z_V} = \left| \frac{F_z}{\delta_{z_2}(L-\delta)} \right| \tag{4.46}$$

which, upon using Eq. (4.45), gives

$$k_{z_V} = \frac{EwBd^3}{6Lw(-L+L_1+\delta)^2 + BL_1[6L^2 - 6L(L_1+\delta) + L_1(2L_1+3\delta)]}. \tag{4.47}$$

Note the strong dependence of k_{z_V} on the thickness and length of the cantilever. The ratio of the angular and linear spring constants, $\zeta_{z_V}(x)$, is

$$\zeta_{z_T}(x) = \frac{k_{\theta_{z_T}}(x)}{k_{z_T}}. \tag{4.48}$$

Inserting Eqs. (4.42) and (4.47) into Eq. (4.48) gives

$$\zeta_{z_V}(x) = \frac{6Lw(-L+L_1+\delta)^2 + BL_1[6L^2 - 6L(L_1+\delta) + L_1(2L_1+3\delta)]}{3[B(2L-L_1)L_1 + 4Lw(x-L_1)]}. \tag{4.49}$$

4.4.6
Numerical Examples

The slope of the V-shaped cantilever, $\theta_{z_V}(x)$, is shown in Figure 4.8 as a function of the distance x from its base for $F_z = 10$ (solid line), 20 (dashed line), and 30 nN (dotted line). Here, the angular spring constant, $k_{\theta_{z_V}}$, is depicted in the figure.

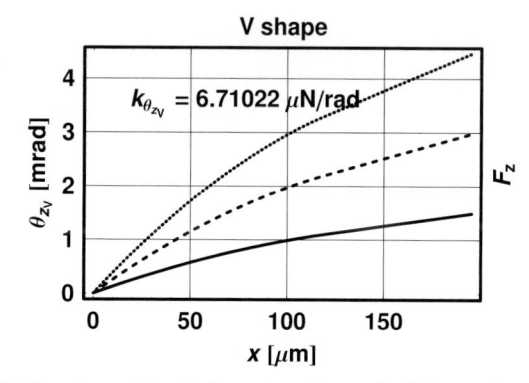

Fig. 4.8 The slope of the V-shaped cantilever, $\theta_{z_V}(x)$, as a function of the distance x from its base for $F_z = 10$ (solid line), 20 (dashed line), and 30 nN (dotted line). The angular spring constant, $k_{\theta_{z_V}}$, is depicted in the figure.

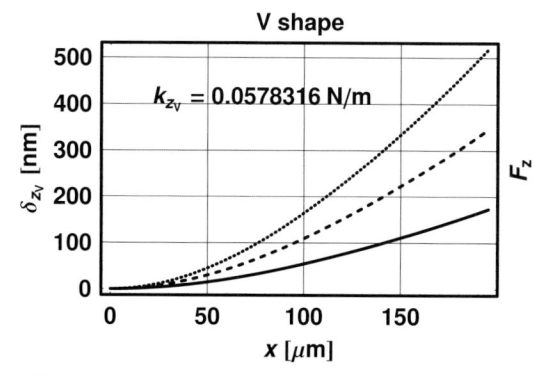

Fig. 4.9 The displacement of the V-shaped cantilever, $\delta_{z_V}(x)$, as a function of the distance x from its base for $F_z = 10$ (solid line), 20 (dashed line), and 30 nN (dotted line). The linear spring constant, k_{z_V}, is depicted in the figure.

The displacement of the V-shaped cantilever, $\delta_{z_V}(x)$, is shown in Figure 4.9 as a function of the distance x from its fixed point, for a tip–sample force acting in the $\hat{z}$ direction. The linear spring constant, k_{z_V}, is depicted in the figure.

Figure 4.10 shows the ratio of the linear and angular spring constants of the V-shaped cantilever, $\zeta_{z_V}(x)$, as a function of the distance x from its base.

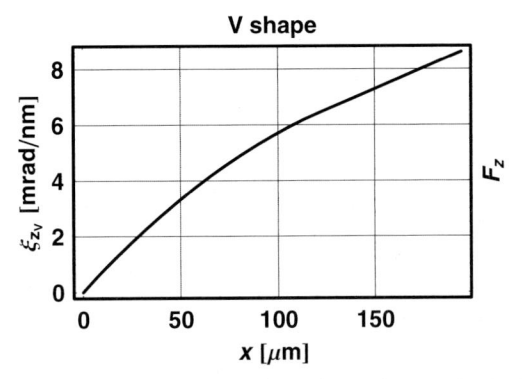

Fig. 4.10 The ratio of the linear and angular spring constants of the triangular cantilever, $\check{\zeta}_{z_V}(x)$, as a function of the distance x from its base.

4.5 Buckling Due to F_x: V Shape

4.5.1 General Equations

In this case, the bending moment, $M_{x_V}(x)$, due to the force F_x, is

$$M_{x_V} = h_t F_x. \tag{4.50}$$

The axial moment of inertia for the range $0 < x < L_1$ is

$$I_{x_1} = \frac{wd^3}{6}. \tag{4.51}$$

The equation of buckling for this range is therefore

$$\frac{\partial^2}{\partial x^2}\delta_{x_1}(x) = \frac{M_{x_V}}{EI_{x_1}}, \tag{4.52}$$

which, using Eqs. (4.50) and (4.51), yields

$$\frac{\partial^2}{\partial x^2}\delta_{x_1}(x) = \frac{6}{Ewd^3}h_t F_x. \tag{4.53}$$

For the range $L_1 < x < L$, the axial moment of inertia is

$$I_{x_2}(x) = \frac{1}{12}\left(1 - \frac{x}{L}\right)Bd^3. \tag{4.54}$$

The equation of buckling for this range is

$$\frac{\partial^2}{\partial x^2}\delta_{x_2}(x) = \frac{M_{x_V}}{EI_{x_2}(x)}. \tag{4.55}$$

Inserting Eqs. (4.50) and (4.54) into Eq. (4.55) yields

$$\frac{\partial^2}{\partial x^2}\delta_{x_2}(x) = \frac{12L}{EBd^3(L-x)}h_t F_x. \qquad (4.56)$$

4.5.2 Slope

The slope, $\theta_{x_1}(x)$, for the range $0 < x < L_1$ is obtained by integrating Eq. (4.56),

$$\theta_{x_1}(x) = \frac{6x}{Ewd^3}h_t F_x. \qquad (4.57)$$

For $L_1 < x < L$, the slope, $\theta_{x_2}(x)$, is obtained by adding the contribution of $\theta_{x_1}(L_1)$ to the integral of Eq. (4.56),

$$\theta_{x_2}(x) = \theta_{x_1}(L_1) + \int_{L_1}^{x} \frac{M_{x_V}(x)}{EI_{x_2}(x)} dx. \qquad (4.58)$$

The resultant integration, using Eqs. (4.50), (4.54), and (4.57), is

$$\theta_{x_2}(x) = 6\left[\frac{L_1}{Ewd^3} + \frac{2L(x-L_1)}{EBd^3(x-L_1)}\log\left(\frac{L-L_1}{L-x}\right)\right]F_x h_t. \qquad (4.59)$$

Note that the slope is a logarithmic function of x and is linear with h_t.

4.5.3 Angular Spring Constant

The angular spring constant, $k_{\theta_{x_V}}(x)$, is the ratio of the force acting on the tip to the resultant slope of the cantilever at the point x,

$$k_{\theta_{x_V}}(x) = \left|\frac{F_x}{\theta_{x_V}(x)}\right|. \qquad (4.60)$$

For the range $L_1 < x < L - \delta$, one can use Eq. (4.59) to obtain

$$k_{\theta_{x_V}}(x) = \frac{EwBd^3}{6Bh_t L_1 + 12h_t Lw\log\left(\frac{L-L_1}{L-x}\right)}. \qquad (4.61)$$

4.5.4 Displacement

For the range $0 < x < L_1$, the displacement, $\delta_{x_1}(x)$, is obtained by integrating Eq. (4.57),

$$\delta_{x_1}(x) = \frac{3x^2}{Ewd^3}h_t F_x. \qquad (4.62)$$

For $L_1 < x < L$, the displacement is obtained by adding the contribution of $\delta_{x_1}(L_1)$ to the integral of Eq. (4.59),

$$\delta_{x_2}(x) = \delta_{x_1}(L_1) + \int_{L_1}^{x} \theta_{x_2}(x)\,dx. \tag{4.63}$$

Using Eqs. (4.59) and (4.62) into Eq. (4.63) yields

$$\delta_{x_2}(x) = \frac{3}{EwBd^3}\left[BL_1(2x - L_1) + 4Lw(x - L_1) - 4Lw(L - x)\log\left(\frac{L-L_1}{L-x}\right)\right]h_tF_x. \tag{4.64}$$

4.5.5
Linear Spring Constant

The cantilever linear spring constant, k_{xV}, is defined as the ratio of the force acting on its tip to the resultant displacement at $x = L - \delta$,

$$k_{xV} = \left|\frac{F_x}{\delta_{x_2}(L-\delta)}\right|, \tag{4.65}$$

which, upon using Eq. (4.64), yields

$$k_{xV} = \frac{(Bd^3Ew)}{3h_t\left[4Lw(L - L_1 - \delta) - BL_1(-2L + L_1 + 2\delta) - 4Lw\delta\log\left(\frac{L-L_1}{\delta}\right)\right]}. \tag{4.66}$$

Note the strong dependence of k_{xV} on the thickness and length of the cantilever and on the value of δ. The ratio of the angular and linear spring constants, $\tilde{\zeta}_{xV}(x)$, is

$$\tilde{\zeta}_{xV}(x) = \frac{k_{\theta_{xV}}(x)}{k_{xV}}. \tag{4.67}$$

Inserting Eqs. (4.61) and (4.66) into Eq. (4.67) gives

$$\tilde{\zeta}_{xV}(x) = \frac{4Lw(L - L_1 - \delta) - BL_1(-2L + L_1 + 2\delta) - 4Lw\delta\log\left(\frac{L-L_1}{\delta}\right)}{2BL_1 + 4Lw\log\left(\frac{L-L_1}{L-x}\right)}. \tag{4.68}$$

4.5.6
Numerical Examples

The slope of the V-shaped cantilever, $\theta_{xV}(x)$, is shown in Figure 4.11 as a function of the distance x from its base for $F_x = 10$ (solid line), 20 (dashed line),

4.5 Buckling Due to Γ_x: V Shape

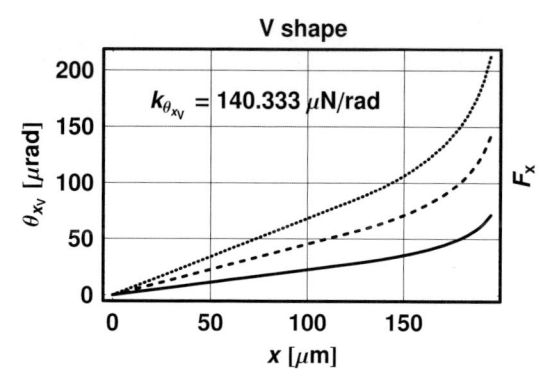

Fig. 4.11 The slope of the V-shaped cantilever, $\theta_{x_V}(x)$, as a function of the distance x from its base for $F_x = 10$ (solid line), 20 (dashed line), and 30 nN (dotted line). The angular spring constant, $k_{\theta_{x_V}}$, is depicted in the figure.

and 30 nN (dotted line). The angular spring constant, $k_{\theta_{x_V}}$, is depicted in the figure.

The displacement of the V-shaped cantilever, $\delta_{x_V}(x)$, is shown in Figure 4.12 as a function of the distance x from its base for $F_x = 10$ (solid line), 20 (dashed line), and 30 nN (dotted line). The linear spring constant, k_{x_V}, is depicted in the figure.

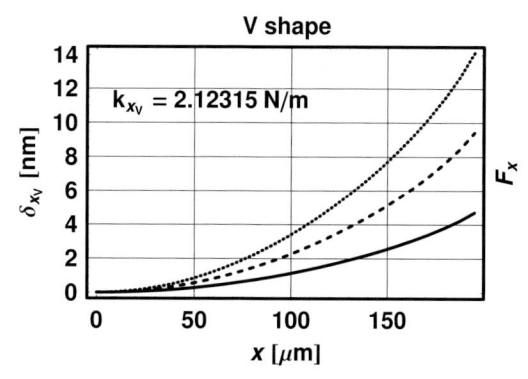

Fig. 4.12 The displacement of the V-shaped cantilever, $\delta_{x_V}(x)$, as a function of the distance x from its base for $F_x = 10$ (solid line), 20 (dashed line), and 30 nN (dotted line). The linear spring constant, k_{x_V}, is depicted in the figure.

Figure 4.13 shows the ratio of the linear to the angular spring constants of the triangular cantilever, $\tilde{\zeta}_{x_V}(x)$, as a function of the distance x from its base.

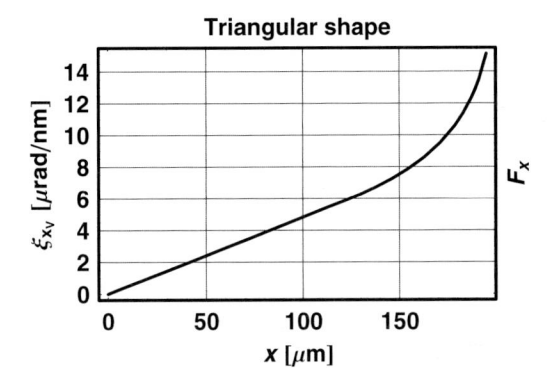

Fig. 4.13 The ratio of the linear to the angular spring constants of the triangular cantilever, $\xi_{xy}(x)$, as a function of the distance x from its base.

4.6 Vibrations

4.6.1 Resonance Frequencies

It is important to know the resonance frequency of a given cantilever, since it sets the upper limit to the scanning speed of an atomic force microscope. In the following discussions we neglect the influence of the mass of the tip and of any cantilever metal coating on the resonance frequency of the cantilever. The time-dependent equation of motion of the V-shaped cantilever, $\delta_{zv}(x,t)$, neglecting damping effects, is given by the differential equation

$$\frac{\partial^2}{\partial x^2}\left[EI_{zv}(x)\frac{\partial^2}{\partial x^2}\delta_z(x,t)\right]+\rho A_V(x)\frac{\partial^2}{\partial t^2}\delta_{zv}(x,t)=0, \tag{4.69}$$

subject to the boundary conditions at the two ends of the cantilever. Here, t denotes time, ρ is the mass density of the cantilever material, I_{zv} and E its axial moment of inertia and Young's modulus, respectively, and $A_V(x)$ its cross-section area. The solution is conveniently obtained using the variational method that equates the kinetic and potential energies of the cantilever. To that end, we construct two matrices, $a_{1_{n,m}}$ and $a_{2_{n,m}}$, where $(n,m) = 1, 2,$ and 3,

$$a_{1_{n,m}} = E\int_0^{L_1}\frac{1}{6}wd^3\partial_{\{x,2\}}x^{n+1}\partial_{\{x,2\}}x^{m+1}\,dx, \tag{4.70}$$

and

$$a_{2_{n,m}} = E\int_{L_1}^{L}\frac{1}{12}Bd^3(1-x/L)\partial_{\{x,2\}}x^{n+1}\partial_{\{x,2\}}x^{m+1}\,dx, \tag{4.71}$$

and add these two,

$$a_{n,m} = a_{1_{n,m}} + a_{2_{n,m}}. \tag{4.72}$$

The value of the matrix $a_{n,m}$ is given in Table 4.2.

Table 4.2 The values of the matrix $a_{n,m}$, multiplied by 10^{16}, where $(n, m) = 1, 2,$ and 3.

$$\begin{pmatrix} 1.0023 & 0.000\,249\,175 & 5.605\,67 \times 10^{-8} \\ 0.000\,249\,175 & 8.408\,51 \times 10^{-8} & 0 \\ 5.605\,67 \times 10^{-8} & 0 & 0 \end{pmatrix}$$

Next, we construct the matrices $b_{1_{n,m}}$ and $b_{2_{n,m}}$, where $(n, m) = 1, 2,$ and 3,

$$b_{1_{n,m}} = \rho \int_0^{L_1} 2wd x^{n+1} x^{m+1} \, dx, \tag{4.73}$$

and

$$b_{2_{n,m}} = \rho \int_{L_1}^{L} Bd\left(1 - \frac{x}{L}\right) x^{n+1} x^{m+1} \, dx, \tag{4.74}$$

and add these two,

$$b_{n,m} = b_{1_{n,m}} + b_{2_{n,m}}. \tag{4.75}$$

The value of the matrix $b_{n,m}$ is given in Table 4.3.

Table 4.3 The value of the matrix $b_{n,m}$ multiplied by 10^{27}, where $(n, m) = 1, 2,$ and 3.

$$\begin{pmatrix} 2.751\,56 & 0.000\,405\,112 & 6.211\,16 \times 10^{-8} \\ 0.000\,405\,112 & 6.211\,16 \times 10^{-8} & 0 \\ 6.211\,16 \times 10^{-8} & 0 & 0 \end{pmatrix}$$

Finally, we obtain the angular resonance frequencies of the cantilever, $\omega_i = 2\pi f_i$, by calculating the roots of the determinant

$$[a_{n,m} - \omega_i^2 b_{n,m}] = 0. \tag{4.76}$$

The first three resonance frequencies, depicted in Table 4.4, show that they are not multiples of each other.

Table 4.4 The first three resonance frequencies of the V-shaped cantilever.

$$f_1 = 24.805\,9\,\text{kHz}$$
$$f_2 = 139.971\,\text{kHz}$$
$$f_3 = 510.982\,\text{kHz}$$

4.6.2
Characteristic Functions

The shape of the vibrating cantilever is obtained from the three matrices, m_i, defined by Ref. [3]

$$m_i = a_{n,m} - \omega_i^2 b_{n,m}. \tag{4.77}$$

The vectors that form the basis for the null space of the matrix m_i (see Table 4.5) serve as the coefficients of the polynomials, V_i, that describe the shape of the free-vibrating cantilever, $\delta_{z_i}(x)$,

$$\delta_{z_i}(x) = V_{i,1} x^2 + V_{i,2} x^3 + V_{i,3} x^4. \tag{4.78}$$

Table 4.5 The vectors that form the basis for the null space of the matrices m_i, for $i = 1, 2,$ and 3.

$$\begin{pmatrix} 2.476\,86 \times 10^{-7} x^2 - 0.000\,820\,266 x^3 + 1x^4 \\ 4.0381 \times 10^{-8} x^2 - 0.000\,431\,064 x^3 + 1x^4 \\ 1.681\,72 \times 10^{-8} x^2 - 0.000\,267\,393 x^3 + 1x^4 \end{pmatrix}$$

The resultant shape of the free-vibrating cantilever, $\delta_{z_i}(x)$, where $i = 1, 2,$ and 3, is shown in Figure 4.14. Note that the curve belonging to the lowest frequency is scaled down by a factor of 10.

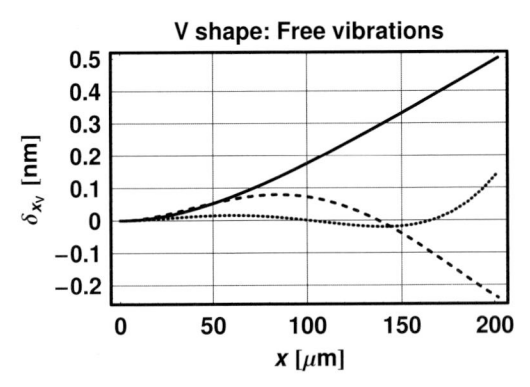

Fig. 4.14 The characteristic functions of the first three harmonics for the V-shaped cantilever.

The solid, dashed, and dotted lines in Figure 4.14 refer to cantilever oscillations at ω_1, ω_2, and ω_3, respectively. Note that a cantilever can obtain a shape which is a linear combination of different characteristic functions, depending on the initial conditions and the driving force. For all practical purposes, however, for as long as the cantilever is far away from a surface and its amplitude of vibration is much smaller than its length, its motion can be considered to be linear in distance and driving force.

▶ Exercises for Chapter 4

1. Using the code in this chapter, reproduce the numerical values quoted for the variety of cantilever materials and geometries in Refs. [1] to [3].
2. Calculate and code the twisting spring constant of a V-shaped cantilever.

References

1. J. E. Sader and L. White, "Theoretical analysis of the static deflection of plates for atomic force microscopy," J. Appl. Phys. **74**, 1 (1993).
2. R. J. Warmack, X.-Y. Zheng, T. Thundat, and D. P. Allison, "Friction effects in the deflection of atomic force microscope cantilevers," Rev. Sci. Instrum. **65**, 394 (1994).
3. G. Y. Chen, R. J. Warmack, T. Thundat, and D. P. Allison, "Resonance response of scanning force microscopy cantilevers," Rev. Sci. Instrum. **65**, 2532 (1994).
4. C. A Clifford and M. P. Seah, "The determination of atomic force microscope cantilever spring constants via dimensional methods for nanomechanical analysis," Nanotechnology **16**, 1666 (2005).

5
Tip–Sample Adhesion

Highlights

1. Johnson–Kendall–Roberts (JKR) adhesion model
2. Indentation and hysteresis loop

Abstract

The interaction between the tip of an atomic force microscope and the surface of a sample is modeled as a combination of adhesion and Lennard–Jones forces. Analytical expressions relating the indentation, d, the contact radius, r, and the contact force, F, are presented and plotted for typical experimental values. Maximum permissible values for the parameters characterizing the adhesion are also calculated and plotted. The contribution of indentation and adhesion forces to the contact pressure and its distribution across the contact area are demonstrated. Finally, the hysteresis loop describing the total force as a function of the tip–sample distance on approach and retraction is presented and discussed.

5.1
Introduction

The interaction between the tip and the sample of an atomic force microscope can involve a variety of forces derived from electric, magnetic, and atomic interactions for the noncontact regime, and from atomic, indentation, adhesion, and capillary interactions for the contact regime. In this chapter we limit the interactions for the noncontact regime to the Lennard–Jones force, and for the contact regime to the indentation and adhesion forces. The analytical expressions of these forces, which are relatively simple, have been widely employed in the literature. It is important to note, however, that the indentation and adhesion forces were developed for macroscopic bodies containing a large number of atoms. For microscopic bodies, the theory is only approximate, and has to be viewed as a general guide to what might be expected to happen when only a few atoms belonging to the tip and sample interact. Nevertheless, the material presented here should be useful in describing the main features associated with the tip–sample interaction, forming a basis that the reader can later expand to include more refined models and other forces.

Figure 5.1 shows two spheres whose curvature matches those of the apex of the tip and possibly that of the surface of a sample, R_{tip} and R_{sample}, respectively. Note that R_{sample} will usually be infinite. Applying an external, downward-pointing force, F_{ex}, to the tip will push it into the sample, forming an indentation whose depth and radius are denoted by d and r, respectively. When the external force is pointing upward, the tip is pulled out of the sample, and for a limited range, a neck connecting these two together will be formed. The neck will have a maximum height and radius denoted by d_s and r_s respectively. For a larger force, the neck can no longer be supported by adhesion forces, whereupon it will break and the two spheres will be separated. The total tip–sample force consists of the sum of Lennard–Jones, indentation, and adhesion forces. This total force, which is a function of the tip–sample separation, s, yields a hysteresis loop because the adhesion is a double-valued function of s. This chapter presents analytical solutions to the functional dependence of the different parameters involved in the tip–sample interaction together with numerical examples.

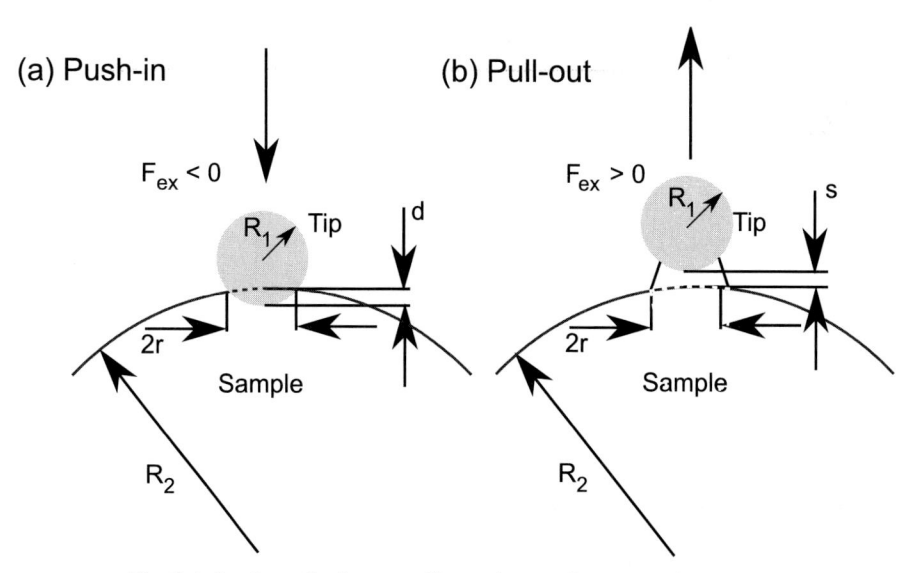

Fig. 5.1 A schematic diagram of two spheres whose curvature represents those of the AFM tip and of the sample, R_1 and R_2, respectively. The tip–sample interactions for (a) approaching and (b) retracting cases are depicted.

The parameters associated with the tip–sample indentation, adhesion, and Lennard–Jones forces used in this chapter are given in Table 5.1. Here, E_i and v_i are Young's modulus and Poisson's ratio of the tip and sample material, respectively, σ is a typical interatomic distance, and A_H is Hamaker's constant.

Table 5.1 The parameters associated with the tip–sample adhesion and Lennard–Jones potential.

$$R_{tip} = 100\,\text{nm}$$
$$R_{sample} = 100\,\mu\text{m}$$
$$E_{tip} = 179\,\text{GN/m}^2$$
$$E_{sample} = 1\,\text{GN/m}^2$$
$$v_{tip} = 0.28\,\mu\text{m}$$
$$v_{sample} = 0.3\,\mu\text{m}$$
$$\sigma = 0.4125\,\text{nm}$$
$$A_H = 0.0135\,\text{aJ}$$

The effective tip–sample radius of curvature, R, is defined by the radii of curvature of the tip and sample, R_{tip} and R_{sample}, respectively,

$$R = \frac{1}{1/R_{tip} + 1/R_{sample}}. \tag{5.1}$$

The Derjaguin approximation for the adhesion force, F_{ad}, acting between two spheres separated by a distance $s \ll R$ is

$$F_{ad} = 2\pi RW(s). \tag{5.2}$$

Here, W is the energy per unit area of two flat surfaces separated by the same distance s. The surface energy, γ, which is the energy required to remove two flat surfaces from contact to infinity is

$$\gamma = \frac{1}{2}W(0). \tag{5.3}$$

The value of γ, in terms of Hamaker's constant, is

$$\gamma = \frac{A_H}{24\pi D_0^2}, \tag{5.4}$$

where D_0 can be approximated by

$$D_0 = \frac{\sigma}{2.5} \tag{5.5}$$

using a typical interatomic distance, σ. The numerical values of D_0, γ, and $W_{1,2}$ are given in Table 5.2.

The Hertzian adhesion model that treats rigid bodies has been extended to deformable ones by Johnson, Kendall, and Roberts (JKR). The JKR model will

Table 5.2 The numerical values of D_0, γ, and $W_{1,2}$.

$D_0 = 0.165\,\text{nm}$
$\gamma = 6.576\,65\,\text{mJ}/\text{m}^2$
$W_{1,2} = 13.1533\,\text{mJ}/\text{m}^2$

be used to calculate the functional dependence of the tip–sample indentation, radius of contact, and contact force on each other. In this model, the elastic moduli, κ, of the two spheres representing the tip and sample are

$$\kappa_{\text{tip}} = \frac{1 - v_{\text{tip}}^2}{\pi E_{\text{tip}}}, \tag{5.6}$$

for the tip, and

$$\kappa_{\text{sample}} = \frac{1 - v_{\text{sample}}^2}{\pi E_{\text{sample}}}, \tag{5.7}$$

for the sample. The effective elastic modulus of the two spheres, κ_{eff}, in terms of κ_i, is

$$\kappa_{\text{eff}} = \kappa_{\text{tip}} + \kappa_{\text{sample}}. \tag{5.8}$$

The numerical values of κ_{tip}, κ_{sample}, and κ_{eff} are given in Table 5.3.

Table 5.3 The numerical values of the elastic moduli κ_{tip}, κ_{sample}, and κ_{eff}.

$\kappa_{\text{tip}} = 0.001\,638\,85\,\text{m}^2/\text{nN}$
$\kappa_{\text{sample}} = 0.289\,662\,\text{m}^2/\text{nN}$
$\kappa_{\text{eff}} = 0.291\,301\,\text{m}^2/\text{nN}$

5.2 Indentation

The following three sections calculate the tip–sample contact radius as a function of the contact force, $r = r(F)$, the tip–sample indentation as a function of the contact radius, $d = d(r)$, and the indentation as a function of the contact force, $d = d(F)$.

5.2.1 Contact Radius and Contact Force

The contact radius between two spheres, r, as a function of the tip–sample contact force, F, is

$$r^3 = \frac{3\pi}{4} \kappa_{\text{eff}} R \left(F + q + \sqrt{q^2 + 2qF} \right), \tag{5.9}$$

where

$$q = 3\pi R W_{1,2}. \tag{5.10}$$

Alternatively, the contact radius can be expressed by

$$r = \left[\frac{R}{k} \left(F + q + \sqrt{q^2 + 2qF} \right) \right]^{1/3}, \tag{5.11}$$

where

$$k = \frac{4}{3\pi \kappa_{\text{eff}}}. \tag{5.12}$$

For $W_{1,2} = 0$, and in the absence of an external force, the tip will rest on the surface of the sample, while for $W_{1,2} > 0$, it will embed itself inside the sample at such a depth that the attractive adhesion force will equal the repulsive indentation force. Under these conditions, the radius of contact, r_0, is

$$r_0 = \left(\frac{6\pi R^2 W_{1,2}}{k} \right)^{1/3}. \tag{5.13}$$

Note that the depth of penetration of the tip into the sample, relative to the effective radius R, is quite small. One can now apply an external force to pull the tip from inside the sample back to its surface. Pulling the tip farther away from the surface of the sample will result in the creation of a neck connecting the tip with the sample. The neck will keep growing until the adhesion force can no longer sustain it, whereupon the neck will break. The tip–sample attractive force at the point of separation, F_s, is obtained by equating the cubic root in Eq. (5.11) to zero, yielding

$$F_s = -\frac{3}{2} \pi R W_{1,2}. \tag{5.14}$$

The radius of the neck at separation, r_s, is

$$r_s = -\frac{r_0}{4^{1/3}}. \tag{5.15}$$

Table 5.4 shows the numerical values of r_0, r_s and F_s.

Figure 5.2 shows the contact radius, r, as a function of the contact force, F, in the range $\pm F_s$, where the solid and dashed lines refer to the cases where $W_{1,2}$ is finite and zero, respectively.

Consider first the case where $W_{1,2} = 0$, shown by the dashed line in Figure 5.2. Here, the radius of contact is finite only for a tip–sample repulsive

Table 5.4 The numerical values of r_0, r_s, and F_s.

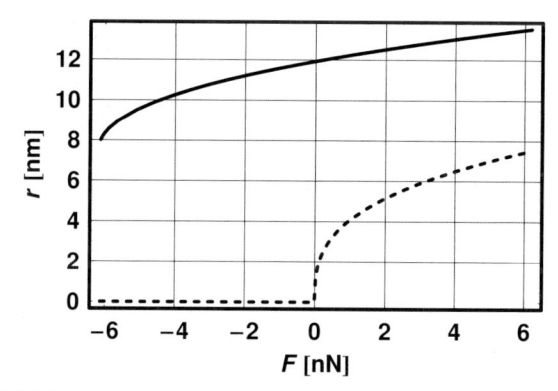

$r_0 = 11.94\,\text{nm}$
$r_s = 7.52\,\text{nm}$
$F_s = -6.19\,\text{nN}$

Fig. 5.2 The contact radius, r, as a function of the contact force, F. The solid and dashed lines refer to the cases where $W_{1,2}$ is finite and zero, respectively.

(positive) force, which is pointing upward in the geometry shown in Figure 5.1. For the case where $W_{1,2} > 0$, however, the radius of contact is finite not only for a positive tip–sample force, as in the previous case, but also for a negative force which is attractive, up to F_s. This is the force required to keep the tip separated from the sample by generating a connecting neck.

5.2.2
Indentation and Contact Radius

We now calculate the tip–sample indentation, d, as a function of the radius of contact, r. The indentation, according to the JKR model, is given by

$$d = -\frac{r^2}{R} + \frac{2}{3}\frac{r_0^{3/2}}{R}\sqrt{r}. \tag{5.16}$$

Figure 5.3 shows the indentation, d, as a function of the radius of contact, r, in the range $\pm F_s$, where the solid and dashed lines refer to the cases where $W_{1,2}$ is finite and zero, respectively. The feature observed at $r = r_s$ is associated with the breaking of the neck connecting the tip with the sample.

Note that a finite $W_{1,2}$ increases the indentation for a given radius of contact. Note also that the indentation can only be negative for a vanishing $W_{1,2}$, whereas it can attain a positive value for a finite $W_{1,2}$ in the form of a neck.

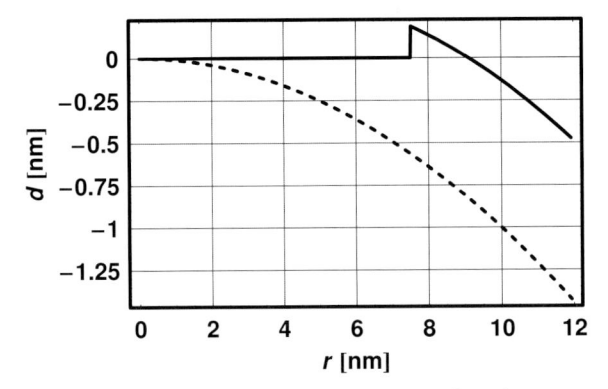

Fig. 5.3 The indentation, d, as a function of the radius of contact, r. The solid and dashed lines refer to the cases where $W_{1,2}$ is finite and zero, respectively.

For the case where $W_{1,2}$ is finite, the radius of contact at which the indentation vanishes, namely, when the height of the neck is zero, will be denoted by r_1. The numerical values of r_1, d_0, and d_s are given in Table 5.5.

Table 5.5 The numerical values of r_1, d_0, and d_s.

$r_1 = 9.11$ nm
$d_0 = -0.47$ nm
$d_s = 0.19$ nm

Note that d_0, d_s, and $W_{1,2}$ vanish together. The negative value of d_0 means that in the absence of a contact force, the tip embeds itself inside the sample. The positive value of d_s denotes the length of the tip–sample "neck" before it breaks.

5.2.3
Indentation and Contact Force

We can now combine the results obtained in the last two sections and get the dependence of the indentation, d, on the contact force, F, shown in Figure 5.4 for the range $\pm F_s$. Here, the solid and dashed lines refer to the cases where $W_{1,2}$ is finite and zero, respectively. Note the feature observed at $F = F_s$ and $d = d_s$ associated with the breaking of the neck connecting the tip with the sample.

First consider the case where $W_{1,2} = 0$, shown by the dashed line, in which a repulsive (positive) tip–sample force is related to a (negative) indentation, using the geometry shown in Figure 5.1. The dashed line goes to zero abruptly at $F = 0$, at which point the tip rests on the surface of the sample. However,

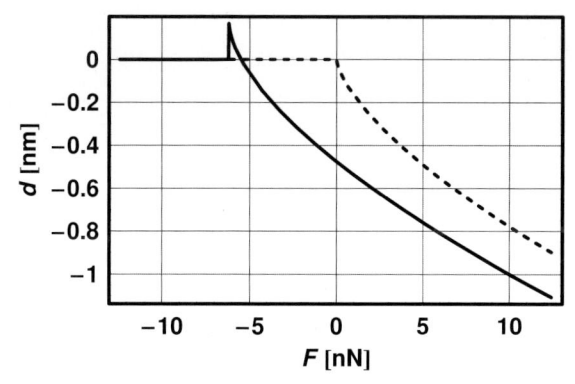

Fig. 5.4 The indentation, d, as a function of the contact force, F. The solid and dashed lines refer to the cases where $W_{1,2}$ is finite and zero, respectively.

for the case where $W_{1,2}$ is finite, the solid line extends to both positive and negative values of F, and is mostly lower than the dashed line. Table 5.6 summarizes the results obtained up to now for the case were $W_{1,2} > 0$.

Table 5.6 Summary of the indentation results.

	r (nm)	d (nm)	F (nN)
$d = d_s$	7.52	0.19	−6.19
$d = 0$	9.11	0	−5.5
$F = 0$	11.94	−0.47	0

5.3 Inverted Functions

It will be useful to invert the functional dependence obtained in the last three sections and calculate the tip–sample contact radius as a function of the contact force, $F = F(r)$, the tip–sample indentation as a function of the contact radius, $r = r(d)$, and the indentation as a function of the contact force, $F = F(d)$.

5.3.1 Contact Force and Contact Radius

The contact force, F, in terms of the radius of contact, r, is obtained by inverting Eq. (5.11),

$$F = \frac{kr^3}{R} - \sqrt{\frac{2kq}{R}} r^{3/2}, \qquad (5.17)$$

which, for a vanishing $W_{1,2}$, becomes

$$F_0 = \frac{kr^3}{R}. \tag{5.18}$$

Figure 5.5 is a plot of the contact force, F, as a function of the contact radius, r, in which the solid and dashed lines refer to the cases where $W_{1,2}$ is finite and zero, respectively.

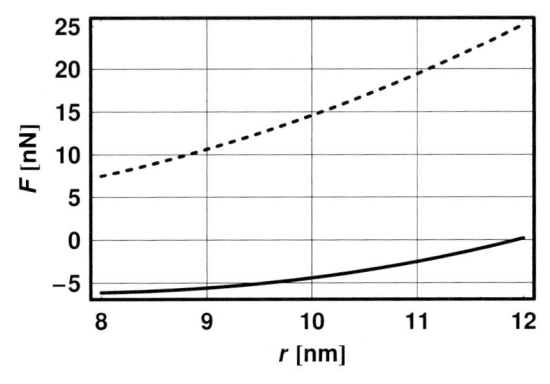

Fig. 5.5 The contact force, F, as a function of the contact radius, r. The solid and dashed lines refer to the cases where $W_{1,2}$ is finite and zero, respectively.

5.3.2
Contact Radius and Indentation

Figure 5.6 shows the contact radius, r, as a function of the indentation, d, obtained by numerically inverting Eq. (5.16). The solid and dashed lines refer to the cases where $W_{1,2}$ is finite and zero, respectively.

5.3.3
Contact Force and Indentation

Now that we have inverted both Eqs (5.11) and (5.16), we can determine the contact force as a function of the indentation. This will be accomplished by using the force as a function of the radius of contact, and the radius of contact as a function of the indentation. Figure 5.7 shows the contact force, F, as a function of the indentation, d, where the solid and dashed lines refer to the cases where $W_{1,2}$ is finite and zero, respectively. The figure shows that a finite value for $W_{1,2}$ is lowering the force for a given indentation. Also, for a vanishing $W_{1,2}$, the range across which the force is finite, is limited to $d < 0$.

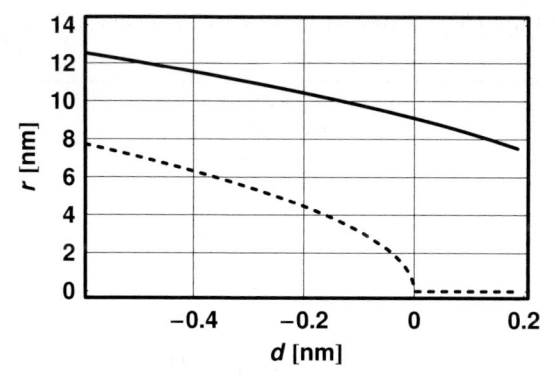

Fig. 5.6 The contact radius, r, as a function of the indentation, d. The solid and dashed lines refer to the cases where $W_{1,2}$ is finite and zero, respectively.

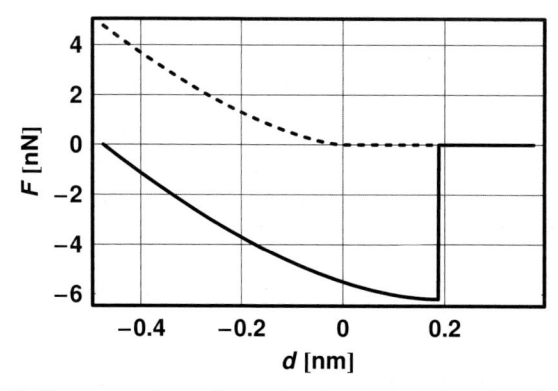

Fig. 5.7 The contact force, F, as a function of the indentation, d. The solid and dashed lines refer to the cases where $W_{1,2}$ is finite and zero, respectively.

5.3.4
Limits of Adhesion Parameters

In the absence of an external force, the tip–sample surface energy will "suck" the tip into the surface until the surface tension equals the indentation force. The maximum value of the adhesion energy, $W_{1,2,\max}$, that will generate a contact radius as large as the effective radius, R, obtained from Eq. (5.11), is given by

$$W_{1,2,\max} = \frac{1}{6\pi} Rk, \tag{5.19}$$

and shown in Figure 5.8.

The maximum value for Hamaker's constant, $A_{H\max}$, for this case,

$$A_{H\max} = 2D_0^2 Rk, \tag{5.20}$$

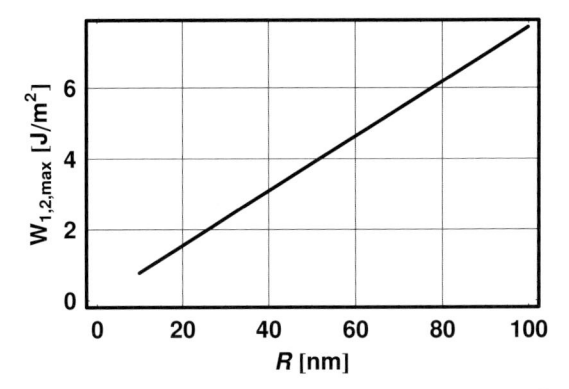

Fig. 5.8 The surface energy, $W_{1,2}$, as a function of the radius, R.

is shown in Figure 5.9 as a function of R.

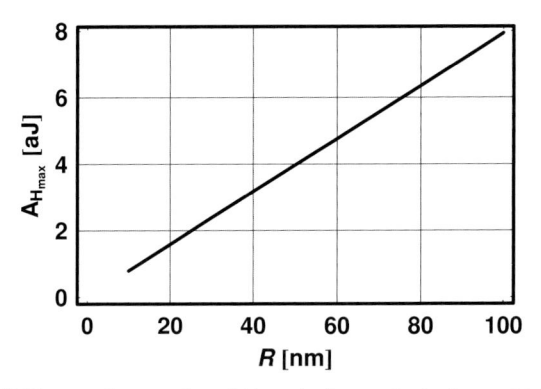

Fig. 5.9 The maximum value of Hamaker's constant, $A_{H_{max}}$, as a function of the radius, R.

5.4
Contact Pressure

The contact pressure, P, across the indented surface is given by

$$P = \frac{3kr}{2\pi R}\sqrt{1-x^2} - \frac{1}{\sqrt{1-x^2}}\sqrt{\frac{3kW_{1,2}}{2\pi r}}, \tag{5.21}$$

where $r/r_0 < x < r/r_0$. Clearly, the second term on the right-hand side diverges for $x = 1$, a region where a more subtle theory is required.

5.4.1
Maximum Contact Pressure

Let us first explore the dependence of the maximum tip–sample contact pressure, P_{max}, realized at $x = 0$, on the radius of contact, r. Using Eq. (5.21), we obtain

$$P_{max} = \frac{3kr}{2\pi R} - \sqrt{\frac{3kW_{1,2}}{2\pi r}}. \qquad (5.22)$$

Figure 5.10 shows P_{max} as a function of the scaled radius of contact, r/r_0, in the range $0.1 < r/r_0 < 2$. The solid and dashed lines refer to the cases where $W_{1,2}$ is finite and zero, respectively.

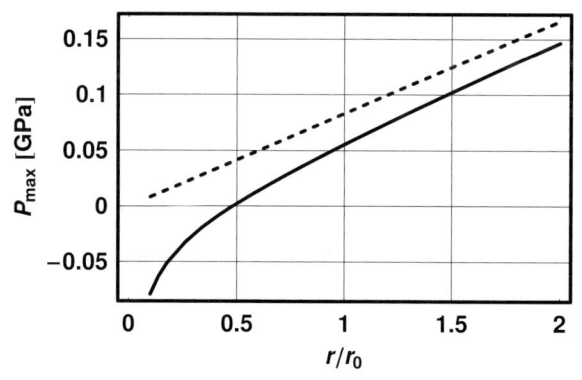

Fig. 5.10 The maximum tip–sample pressure, P_{max}, as a function of the scaled radius of contact, r/r_0. The solid and dashed lines refer to the cases where $W_{1,2}$ is finite and zero, respectively.

One observes that the effect of the adhesion is more prominent for smaller radii. The two contributions on the left-hand side of Eq. (5.22) will be equal for a radius of contact, r_p, given by

$$r_p = \left(\frac{2\pi R^2 W_{1,2}}{3 k} \right)^{1/3}. \qquad (5.23)$$

5.4.2
Distribution of Contact Pressure

The distribution of the contact pressure, P, across the semispherical tip–sample indentation, Eq. (5.21), is shown in Figure 5.11, for a radius of contact given by $r = r_0$. Here, x is in the range $-0.6r_0 < x < 0.6r_0$. The solid and dashed lines refer to the cases where $W_{1,2}$ is finite and zero, respectively.

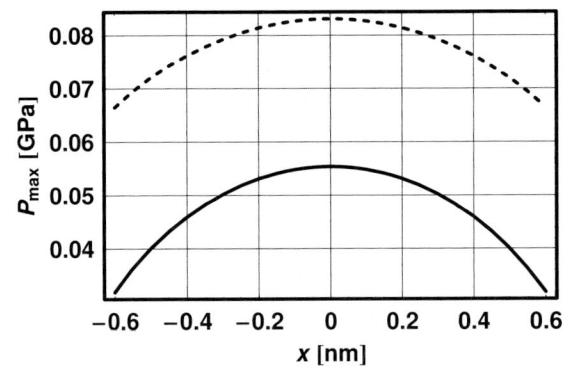

Fig. 5.11 The distribution of the pressure, P, as a function of the position, x, in the range $-0.6r_0 < x < 0.6r_0$. The solid and dashed lines refer to the cases where $W_{1,2}$ is finite and zero, respectively.

5.5
Lennard–Jones Potential

We will need to choose a model for the interaction between the tip and the sample when they are not in contact. We choose the Lennard–Jones model that presents the tip–sample potential energy, W_{LJ}, by

$$W_{LJ} = \frac{A_H R}{6\sigma} \left[\frac{1}{210} \left(\frac{\sigma}{s} \right)^7 - \frac{\sigma}{s} \right], \tag{5.24}$$

where σ is a typical interatomic distance and s is a measure of the tip–sample distance. The factor on the right-hand side is the magnitude of the potential in terms of A_H and R, and the term in the bracket, on the right-hand side, describes the shape of the potential. The Lennard–Jones force, F_{LJ}, is obtained from the potential by taking the negative of its derivative in respect to s,

$$F_{LJ} = -\partial_s W_{LJ}, \tag{5.25}$$

which yields

$$F_{LJ} = \frac{A_H R}{6\sigma} \left(\frac{\sigma^7}{30 s^8} - \frac{\sigma}{s^2} \right). \tag{5.26}$$

Let s_{F_0} and $s_{F_{min}}$ denote the values of s at which the force vanishes and at which it obtains its minimum value, respectively. In the following, it will be convenient to (a) define a shifted tip–sample distance, $d = s + s_{F_0}$, and (b) assume that F_{LJ} vanishes for $d < 0$. Figure 5.12 shows the modified Lennard–Jones force as a function of the shifted tip–sample distance, d. From now on we will refer to the modified Lennard–Jones force as the Lennard–Jones force. Note that for $d > 0$, the force is negative, and by convention, attractive.

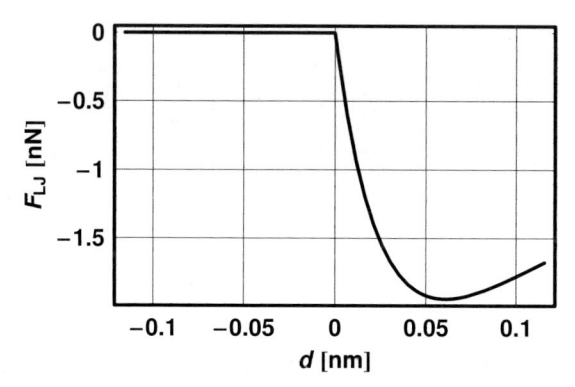

Fig. 5.12 The modified Lennard–Jones force, F_{LJ}, as a function of the shifted tip–sample distance, d.

5.6 Total Force and Indentation

The total tip–sample force, F_{total}, is obtained by combining the indentation, adhesion, and Lennard–Jones forces. The combined force, when plotted as a function of d, exhibits a hysteresis loop because the adhesion is not a single-valued function due to the formation of the neck. It is quite different when the tip approaches the sample from a large distance and is then pushed into it, relative to when the tip is pulled out of the sample until it is separated from it. The discussion will now follow the different parts of the hysteresis loop, after which they will all be combined together.

5.6.1 Push-in Region

For the push-in region, we start by positioning the tip far away from the surface of the sample, after which it is moved toward the sample until it "touches" its surface, and finally, it is pushed into the sample, indenting it.

5.6.2 Push-in Region in the Absence of Adhesion

In the absence of adhesion, the tip indents the sample only for $d < 0$, where the force is given by Eq. (5.18), while for $d > 0$, the force is due to the Lennard–Jones force, Eq. (5.26), as shown in Figure 5.13. As expected, the force is attractive for $d > 0$ and repulsive for $d < 0$. Note the abrupt change in the derivative of the force as it "touches" the surface of the sample. Also note that the repulsive part of the force will have a larger slope for stiffer tip and sample materials.

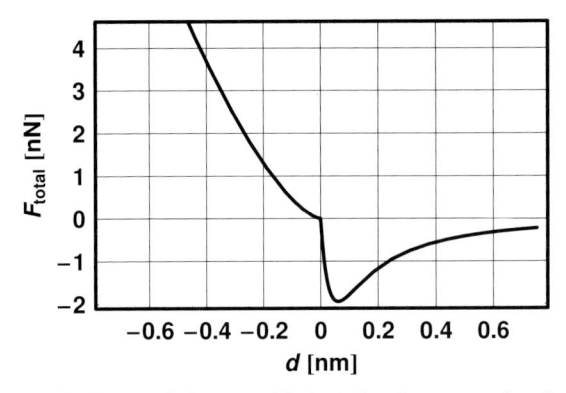

Fig. 5.13 The Lennard–Jones and indentation forces as a function of d in the absence of adhesion.

5.6.3
Push-in Region in the Presence of Adhesion

In this case, once the tip "touches" the surface, the tip–sample surface energy will "suck" it into the sample until the indentation force will equal the adhesion force, both given by Eq. (5.17). Figure 5.14 shows the combined Lennard–Jones, indentation, and adhesion forces as a function of d. In contrast to Figure 5.13, here the force undergoes an abrupt change as the tip "touches" the surface of the sample and is then "sucked" into it. At this point there is still a finite attractive force that will decrease as the indentation force grows, until it equals the adhesion force. Pushing the tip deeper into the sample will result in a repulsive force. As before, there is an abrupt change in the derivative of the force as it "touches" the surface of the sample, and the repulsive part of the force will have a larger slope for stiffer tip and sample materials.

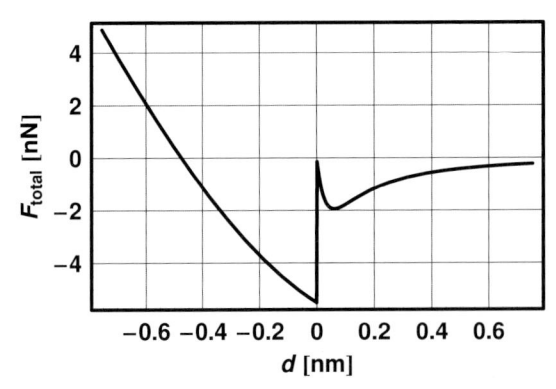

Fig. 5.14 The combined Lennard–Jones, indentation, and adhesion forces as a function of d, in the presence of adhesion.

5.6.4
Pull-out Region

For the pull-out region, we start with a tip that is already embedded inside the sample, and move it toward its surface, after which it is moved away from the sample. A neck will develop between the tip and the sample once d becomes positive, until it can no longer be sustained by the surface energy, upon which it will break. At this point the attractive Lennard–Jones potential will take over.

5.6.5
Pull-out Region in the Absence of Adhesion

In the absence of adhesion, the tip indents the sample for negative values of d, and the force follows the expressions obtained before. For positive values of d, the force will be solely due to the attractive Lennard–Jones potential. The plot of the force in this case will be identical to that of Figure 5.13, as the adhesion is a single-valued function for a vanishing $W_{1,2}$.

5.6.6
Pull-out Region in the Presence of Adhesion

The tip in this case is partially embedded inside the sample, where a repulsive force acts on it. As the tip approaches the surface of the sample, the repulsive force decreases, and at a given point it vanishes after which it becomes attractive. The attractive force is finite not only at $d = 0$ but also up to $d = d_s$, where the tip–sample neck finally breaks. At this point the force will be due only to the attractive Lennard–Jones potential. The plot of the force in this case will be identical to that of Figure 5.14, since the adhesion is now a double-valued function for a finite $W_{1,2}$.

5.6.7
Hysteresis Loop

Figure 5.15 shows the hysteresis loop describing the complete push-in (solid line) and pull-out (dashed line) processes experienced by the tip in a full cycle, which is a combination of Figures 5.13 and 5.14. Note that this hysteresis loop cannot be directly viewed by experiments with an atomic force microscope because the tip is mounted on a cantilever that has a finite spring constant. We have not taken into account the elastic force associated with the bending of the cantilever to which the tip is attached, a topic that will be dealt with in the next chapter.

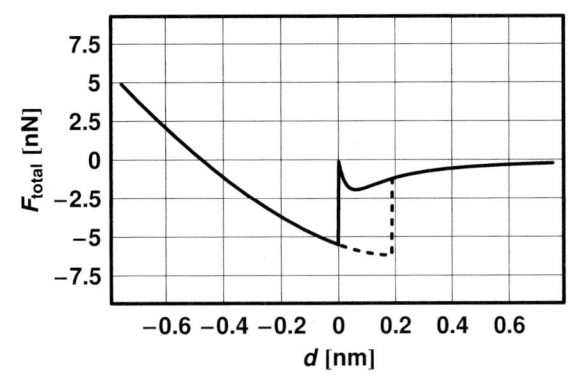

Fig. 5.15 The hysteresis loop describing the complete push-in (solid line) and pull-out (dashed line) processes experienced by the tip in a full cycle.

▶ **Exercises for Chapter 5**

1. Change the sample material and run the code of this chapter using the new parameters.
2. Compare and discuss the current and new figures and, in particular, their respective hysteresis loops.
3. Repeat Problem 1, but this time change Hamaker's constant from its value A_H to $A_H/10$.

References

1 K. L. Johnson, K. Kendall, and A. D. Roberts, "Surface energy and the contact of elastic solids," Proc. R. Soc. London A **324**, 301 (1971).
2 K. Kendall, "The adhesion and surface energy of elastic solids," J. Phys. D: Appl. Phys. **4**, 1186 (1971).
3 J. Israelachvili, *Intermolecular and Surface Forces*, 2nd Edition, Academic Press, San Diego, 1992.

6
Tip–Sample Force Curve

■ Highlights

1. Hysteresis loop of tip–sample approach and retraction
2. Snap-in and snap-out positions
3. Hysteresis loop area and Hamaker's constant

Abstract

The interaction between the tip of an atomic force microscope and a sample is modeled by using the elastic properties of a cantilever and the Lennard–Jones potential, neglecting adhesion and indentation forces. Analytical and numerical solutions for the resultant energy, force, and force derivative, together with their extrema, are derived. The hysteresis loop observed on approach and retraction of the tip at and from the sample is presented and total energy considerations are discussed. It is shown that the area enclosed by the hysteresis loop is proportional to the square of Hamaker's constant.

6.1
Introduction

An atomic force microscope consists of a cantilever with a sharp tip at its end that is brought into close proximity to the surface of a sample. One can fix the position of the sample and mount the cantilever base on an x–y–z scanner, a configuration used in "stand-alone" systems aimed at probing large samples. Alternatively, one can mount the sample on a scanner and fix the position of the base of the cantilever, as is conveniently employed in the design of most atomic force microscopes that probe small samples. We will be using the first geometry, which is shown schematically in Figure 6.1, noting that the modeling will produce the same results if we use the other geometry. The stationary sample is assumed to be at "ground" level, positioned at $g = 0$. The base of the cantilever, at a position u, is mounted on a bimorph that controls its height relative to the sample. The presence of a tip–sample force will bend the cantilever, deflecting the position of the tip, s, from its original position, u. The deflection of the cantilever, d, will therefore be given by $d = s - u$. In a real experiment, the tip–sample force may involve contributions from atomic, adhesion, and indentation forces, to name only a few. However, the salient features of an atomic force microscope can be obtained quite accurately by modeling the tip–sample interaction using only a Lennard–Jones potential.

The main topic discussed in this chapter will concern the deflection of the cantilever as a function of the bimorph position, namely $d = d(u)$. Note that the value of u is computer controlled and the value of d is measured using optical deflection or interferometric methods. Clearly, the tip–sample distance, s, is determined by both u and d. For convenience, however, the modeling will use s as a free parameter so that the plot of $d = d(u)$ will be accomplished by using a parametric plot of $d(s)$ against $u(s)$ for a particular range of s values. The operation of the atomic force microscope, in the so-called contact mode, will usually start by having the base of the cantilever placed far enough from the sample such that the tip does not interact with it. A mechanical motor will then lower the base, decreasing the tip–sample distance, until a slight cantilever deflection is monitored by the AFM. At this point, the AFM motor will stop, and the computer-controlled bimorph will take over, moving the cantilever base closer to the sample. The downward deflection of the cantilever is continuously monitored as its tip approaches the sample. At a particular point, denoted by u_{in}, the tip will snap into the surface of the sample. Moving the base of the cantilever even closer to the surface of the sample will decrease the cantilever deflection, which eventually will change direction. When the base of the cantilever is pulled away from the sample, the deflection of the cantilever again changes directions, and at a given point, denoted by u_{out}, the tip will snap out from the surface of the sample. The cantilever at this point is still bent downward under the influence of the tip–sample force. However, when the base of the cantilever is retracted far enough from the sample, the deflection of the cantilever will vanish.

Tracing this deflection as a function of the bimorph position yields a force curve resembling a hysteresis loop that characterizes the operation of the contact-mode atomic force microscope. Analysis of the cantilever, tip and sample interactions shed light on the significance of the "snap-in" and "snap-out" points. The calculation of the area enclosed under the hysteresis loop indicates that one can measure Hamaker's constant from experimentally obtained force curves. By augmenting the modeling presented in this chapter so that it also includes adhesion and indentation forces, one can obtain information regarding the "stickiness" and "softness" of the sample, respectively.

The parameters of the system, which are typical to those of a commercial AFM, are given in Table 6.1.

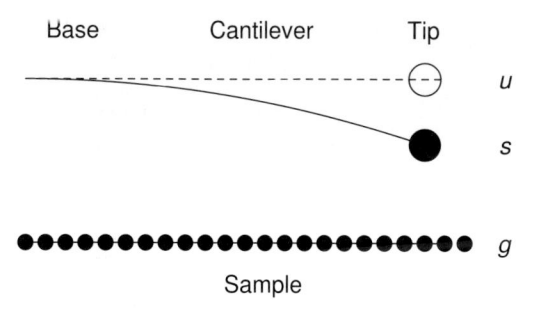

Fig. 6.1 Schematic diagram of the cantilever, tip, and sample. Here, the surface of the sample, which is stationary, is positioned at g. The position of the base of the cantilever is at u, while the position of the tip is at s.

Table 6.1 The parameters of the systems.

$R = 10\,\text{nm}$
$\sigma = 0.35\,\text{nm}$
$A_H = 0.425\,\text{aJ}$
$k = 0.5\,\text{N/m}$

6.2 Tip–Sample Interaction

6.2.1 Lennard–Jones Potential

Of the several forces a tip may encounter when placed in the vicinity of a sample, we consider only the one derived from the Lennard–Jones potential, W, given by

$$W = \frac{A_H R}{6\sigma}\left[\frac{1}{210}\left(\frac{\sigma}{s}\right)^7 - \frac{\sigma}{s}\right]. \tag{6.1}$$

The first term on the right-hand side is the magnitude of the potential in terms of Hamaker's constant, A_H, and tip radius, R. The second term on the right-hand side describes the shape of the potential in terms of the tip–sample distance, s, and a typical atomic distance, σ. The potential vanishes at a distance, s_{W_0},

$$s_{W_0} = \frac{1}{210^{1/6}}\sigma, \tag{6.2}$$

and obtains its minimum value, $W_{\min}$,

$$W_{\min} = -\frac{30^{1/6}}{7}\frac{A_H R}{\sigma}, \tag{6.3}$$

at a distance, $s_{W_{\min}}$,

6 Tip–Sample Force Curve

$$s_{W_{min}} = \frac{1}{30^{1/6}}\sigma. \tag{6.4}$$

Table 6.2 gives the values of s_{W_0}, $s_{W_{min}}$, and W_{min} for the parameters presented in Table 6.1.

Table 6.2 The values of s_{W_0}, $s_{W_{min}}$, and W_{min} for the parameters presented in Table 6.1.

$s_{W_0} = 0.143\,559$ nm
$s_{W_{min}} = 0.198\,555$ nm
$W_{min} = -3.0578$ aJ

Figure 6.2 shows the tip–sample Lennard–Jones potential, W, as a function of the tip–sample distance, s. Note that the potential, in the region described by Figure 6.2, is attractive and by convention negative.

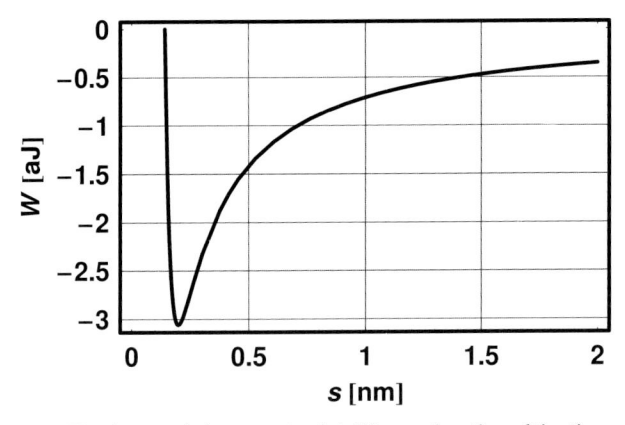

Fig. 6.2 The Lennard–Jones potential, W, as a function of the tip–sample distance, s.

6.2.2
Lennard–Jones Force

The Lennard–Jones force, F, is obtained from the potential by taking the negative of its derivative with respect to s,

$$F = \frac{A_H R}{6\sigma^2}\left(\frac{\sigma^2}{s^2} - \frac{1}{30}\frac{\sigma^8}{s^8}\right). \tag{6.5}$$

The force vanishes at s_{F_0},

$$s_{F_0} = \frac{1}{30^{1/6}}\sigma, \tag{6.6}$$

and obtains its minimum value, F_{min},

$$F_{min} = -\frac{1}{8}\left(\frac{15}{2}\right)^{1/3}\frac{A_H R}{\sigma^2}, \qquad (6.7)$$

at $s_{F_{min}}$,

$$s_{F_{min}} = \left(\frac{2}{15}\right)^{1/6}\sigma. \qquad (6.8)$$

Table 6.3 gives the value of s_{F_0}, $s_{F_{min}}$, and F_{min} for the parameters presented in Table 6.1.

Table 6.3 The value of s_{F_0}, $s_{F_{min}}$, and F_{min} for the parameters presented in Table 6.1.

$s_{F_0} = 0.198\,555$ nm
$s_{F_{min}} = 0.250\,164$ nm
$F_{min} = -8.488\,87$ nN

Figure 6.3 shows the Lennard–Jones force, F, as a function of the tip–sample distance, s. The tip–sample force, in the region described by Figure 6.3, is attractive and by convention negative.

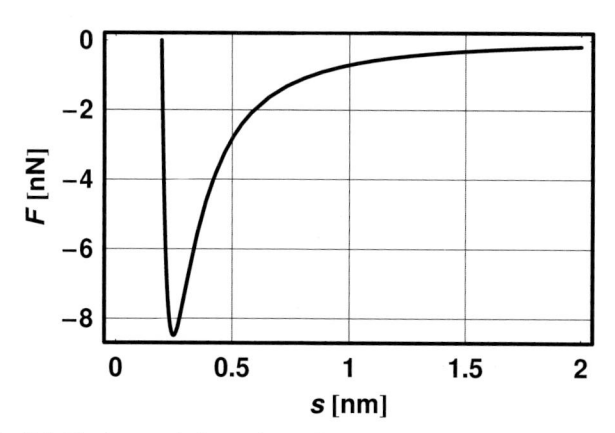

Fig. 6.3 The Lennard–Jones force, F, as a function of the tip–sample distance, s.

6.2.3
Lennard–Jones Force Derivative

The tip–sample Lennard–Jones force derivative, $F' = \partial F/\partial s$, is an important function because it determines whether the tip will or will not "snap" into the sample on approach. F' is obtained from the force F by taking its derivative with respect to s,

$$F' = \frac{A_H R}{3\sigma^3}\left(\frac{\sigma^3}{s^3} - \frac{2}{15}\frac{\sigma^9}{s^9}\right), \qquad (6.9)$$

which vanishes at $s_{F'}$

$$s_{F'_0} = \left(\frac{2}{15}\right)^{1/6}\sigma. \qquad (6.10)$$

F' obtains its maximum value, F'_{max}

$$F'_{max} = \frac{\sqrt{10}}{9}\frac{A_H R}{\sigma^3}, \qquad (6.11)$$

at $s_{F'_{max}}$,

$$s_{F'_{max}} = \left(\frac{2}{5}\right)^{1/6}\sigma. \qquad (6.12)$$

Table 6.4 gives the values of $s_{F'_0}$, $s_{F'_{min}}$, and F'_{min} for the parameters presented in Table 6.1.

Table 6.4 The values of $s_{F'_0}$, $s_{F'_{min}}$, and F'_{min} for the parameters presented in Table 6.1.

$s_{F'_0}$	$= 0.250\,164$ nm
$s_{F'_{max}}$	$= 0.300\,431$ nm
F'_{max}	$= 34.829\,1$ N/m

Figure 6.4 shows the Lennard–Jones force derivative, F', as a function of the tip–sample distance s. Note that the force derivative in the region described in Figure 6.4 is positive, and may obtain values smaller or greater than the spring constant, k, of the cantilever.

6.2.4
Morse Potential

Some calculations appearing in the literature that deal with the tip–sample force involved in scanning tunneling microscopy use a Morse, rather than a Lennard–Jones potential. The Morse force, F_M, is given by

$$F_M = c_1\left[e^{-2c_2(s-s_2)} - 2e^{-c_2(s-s_2)}\right]. \qquad (6.13)$$

It is instructive to find the coefficients c_1 and c_2 such that the position and value of the minimum of this force are the same as those belonging to the

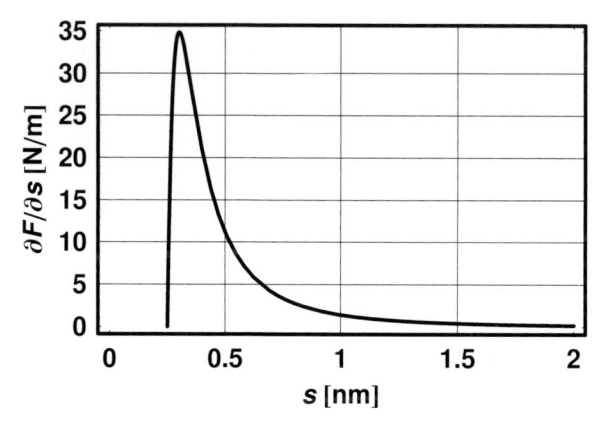

Fig. 6.4 The Lennard–Jones force derivative, F', as a function of the tip–sample distance, s.

Lennard–Jones force. These two coefficients are given in terms of the parameters of the Lennard–Jones potential by

$$c_1 = -\frac{A_H R}{6\sigma^2}\left[\frac{1}{30}\left(\frac{15}{2}\right)^{8/6} - \left(\frac{15}{2}\right)^{1/3}\right], \qquad (6.14)$$

and

$$c_2 = -\frac{\ln(2)}{\sigma}\left[30^{1/6} - \left(\frac{2}{15}\right)^{1/6}\right]. \qquad (6.15)$$

Table 6.5 gives the values of the coefficients c_1 and c_2 of the Morse force.

Table 6.5 The values of the coefficients c_1 and c_2 of the Morse force. The position and value of the minimum of this force are the same as those belonging to the Lennard–Jones force.

$c_1 = 8.48887\,\text{nN}$
$c_2 = 13.4308\,\text{Gm}$

Figure 6.5 shows both the Lennard–Jones (dashed line) and Morse (solid line) forces, F, as a function of the tip–sample distance, s. Note that the range of the Morse force is considerably shorter than that of the Lennard–Jones force.

6.3 Hysteresis Loop

We now discuss the physics associated with the hysteresis loop obtained by plotting the deflection of the cantilever, d, as a function of the position of its

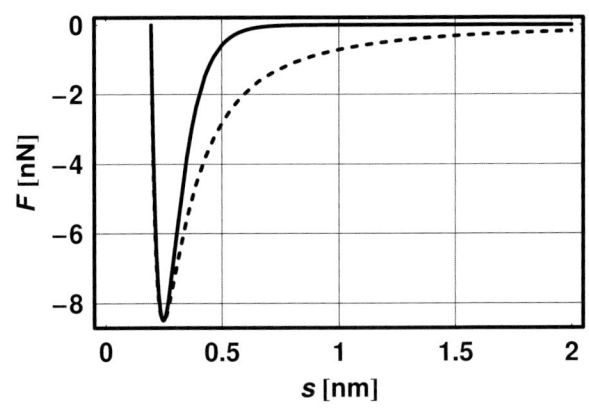

Fig. 6.5 The Morse (solid line) and Lennard–Jones (dashed line) force, F, as a function of the tip–sample distance, s.

base, u. First, the base of the cantilever, which is attached to a bimorph, is moved down until the tip of the cantilever makes contact with the surface of the sample. Then, the base is moved up until the tip looses contact with this surface. We will follow these two motions and find that they produce a hysteresis loop with two distinctive points, one where the tip snaps into the sample and the other one where it snaps out.

6.3.1
Snap-in and Snap-out Points

To obtain a hysteresis loop, the value of the spring constant of the cantilever, k, has to be smaller than the maximum value of the tip–sample force derivative, F'_{max}. To characterize the tip–sample interaction, we plot the deflection of the cantilever, d, as a function of the position of its base, u. Mathematically, it is much simpler to use s as a free parameter and employ a parametric plot of $u(s)$ against $d(s)$, where

$$u = s - \frac{F}{k} \qquad (6.16)$$

and

$$d = \frac{F}{k}. \qquad (6.17)$$

Note that u is controlled by the AFM electronics, while both d and s are determined by the parameters of the system. There are two points where the function d is found to be discontinuous. These are given by the snap-in and snap-out values of s, obtained by solving

$$F'(s_{\text{snap}}) = k. \tag{6.18}$$

Since the Lennard–Jones potential is a seventh-order polynomial, we must sort out the two proper solutions, s_{in} and s_{out}, from which the values of u_{in} and u_{out}, and d_{in} and d_{out}, are calculated. Table 6.6 presents a summary of the numerical values of the six parameters associated with the snap-in and snap-out points. These parameters will be used to describe the hysteresis loop.

Table 6.6 The numerical values of the six parameters associated with the snap-in and snap-out points.

	s (nm)	u (nm)	d (nm)
Snap-in	1.415 02	2.122 54	−0.707 524
Snap-out	0.250 396	17.228	−16.977 6

6.3.2
Calculated Hysteresis Loop

The parametric plot of $u(s)$ against $d(s)$, using s as the free parameter, is shown in Figure 6.6, where the filled disk on the right-hand side denotes the position of the cantilever base far away from the surface of the sample. The function $d(u)$ is observed to be double valued, which does not present a realistic picture of what one expects to observe experimentally during the approach and retraction of the base of the cantilever. To understand the plot shown in Figure 6.6, we (a) dissect it into segments associated with different processes occurring during the cantilever-sample approach and retraction, and (b) redraw it so that it resembles an experimentally measured hysteresis loop. This will be done in the next section where the snap-in and snap-out points are described. Figure 6.6 describes the situation where the bimorph moves the base of the cantilever from a point far away from the sample, denoted in the figure by a filled disk, toward the sample.

6.3.3
Observed Hysteresis Loop

Let us now recalculate the plot shown in Figure 6.6, which will be denoted as the hysteresis loop, as it would appear in an experiment, breaking it into several segments and discussing the physics associated with each one. The first segment of the hysteresis loop, shown in Figure 6.7, describes the situation where the bimorph moves the base of the cantilever from a point far away

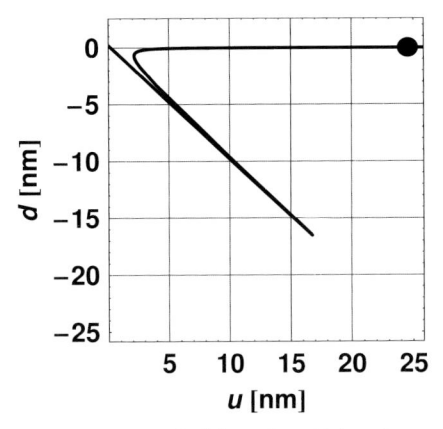

Fig. 6.6 The parametric plot of $u(s)$ against $d(s)$, using s as the free parameter.

from the sample toward its surface. Since at a large distance the tip–sample force can be neglected, the cantilever does not deflect so that $d(u)$ is a constant. The resultant horizontal curve to the left of the filled disk will eventually start dropping down as the tip is drawn toward the sample. The continuous motion of the tip toward the sample will end abruptly at s_{in}, at which point the force derivative equals the spring constant.

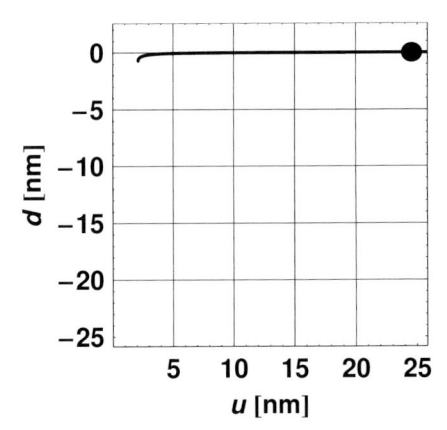

Fig. 6.7 The observed hysteresis loop: from far away to just before the snap-in point.

Once the tip–sample distance s reaches the value s_{in}, the force derivative will overpower the cantilever's spring constant, and the tip will snap into the surface of the sample. The second segment of the hysteresis loop, shown in Figure 6.8, is therefore a straight vertical line, depicted by a downward pointing arrow.

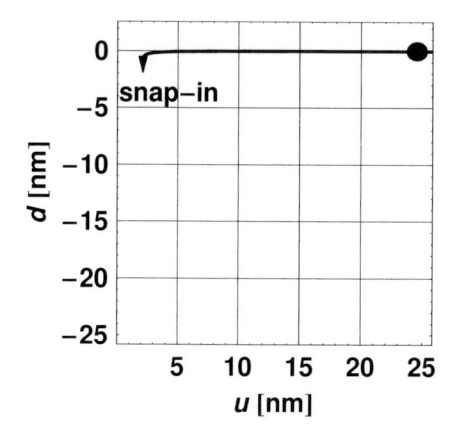

Fig. 6.8 The observed hysteresis loop: from far away to just after the snap-in point.

In Figure 6.8 we plot the calculated position of the head of the arrow, where the tip "touches" the surface of the sample. This position is found by solving $u(s) = u_{in}$. Up to now the cantilever deflected downward because $s < u$. As the bimorph continues to move the base of the cantilever ever closer to the sample, the cantilever deflection will decrease until it vanishes when the Lennard–Jones force vanishes. After passing this point the cantilever will deflect upward, since now $s > u$. The segment of the hysteresis loop describing this process is shown in Figure 6.9 by the straight, diagonal line starting from the head of the arrow. Suppose now that the bimorph starts to retract, pulling up the base of the cantilever. The tip will still be in contact with the sample for as long as the tip–sample force derivative is larger than the restoring force of the cantilever.

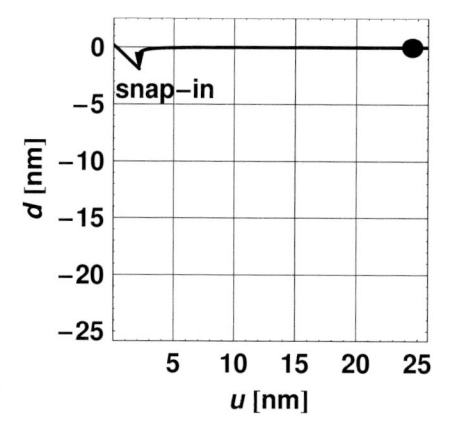

Fig. 6.9 The observed hysteresis loop: from far away to the minimum tip–sample distance.

The fourth segment, shown in Figure 6.10 together with the previous three, is a diagonal line that partially retraces the previous diagonal line. However, the new line continues diagonally downward to a point where the tip will snap out of the sample.

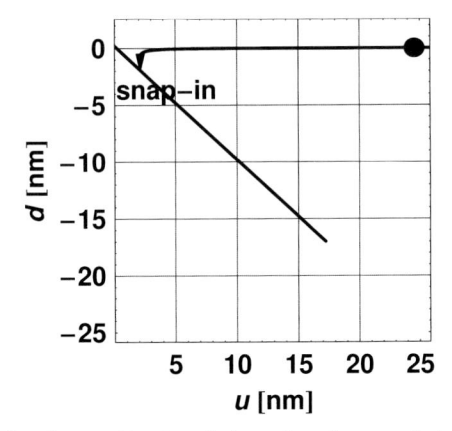

Fig. 6.10 The observed hysteresis loop: from far away to just before the snap-out point.

The snap-out will occur when the tip–sample force derivative decreases below the spring constant. This is described in Figure 6.11 by the arrow pointing upward. Here, again, we have to solve $u(s) = s_{\text{out}}$ to get the appropriate point at which the snap-out process ends. The head of the arrow will now meet the almost horizontal line, as shown in Figure 6.11, and a further retraction of the bimorph will trace this line to the right, while an approach will retrace it to the left. Figure 6.11 describes the observed hysteresis loop obtained by the tip approaching the sample with a subsequent withdrawal.

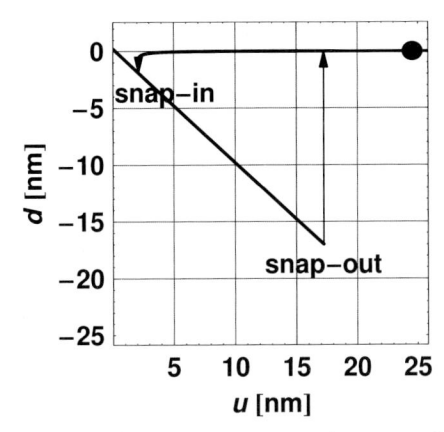

Fig. 6.11 The observed hysteresis loop: from far away to just after the snap-out point.

To better understand the response of the cantilever to the Lennard–Jones potential, we plot in Figure 6.12 the functions $F(s)$, (solid line), $-u(s)$ (dashed line), and $d(s)$ (dotted line). For large values of s, namely, when the tip is far away from the surface of the sample, both the Lennard–Jones force and the deflection of the cantilever are negligible, and their respective curves overlap. Also, u and s, in this range, have a linear relationship, since the tip moves down together with the base of the cantilever. At close proximity to the surface of the sample, however, where s becomes very small, the curves associated with u and s merge. In this range, the three curves exhibit a similar shape because of the smallness of s.

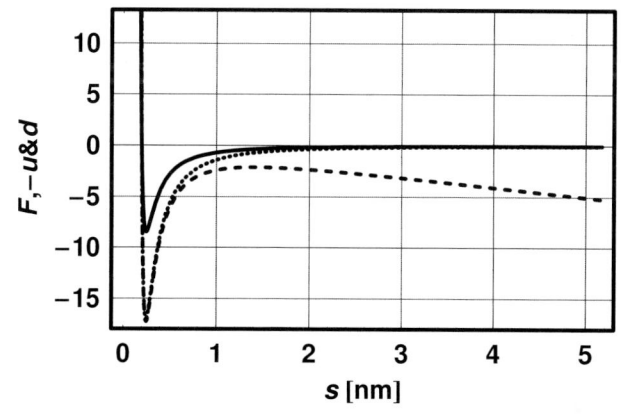

Fig. 6.12 The functions $F(s)$ (solid line), $-u(s)$ (dashed line), and $d(s)$ (dotted line), showing their similar shape.

6.4 Evaluation of Hamaker's Constant

We will now show that one can obtain the value of Hamaker's constant, A_H, from the experimentally observed area enclosed by the hysteresis loop, A_h. To that end, we first evaluate the magnitude of the various parameters associated with the segments shown in Figure 6.11, which are given in Table 6.7.

Table 6.7 The magnitude of the various parameters associated with the segments shown in Figure 6.11.

$s_{out} - s_{in} = -1.16462\,\text{nm}$
$u_{out} - u_{in} = 15.1055\,\text{nm}$
$d_{out} - d_{in} = -16.2701\,\text{nm}$

A good approximation to the area enclosed by the hysteresis loop, shown in Figure 6.11, is readily given by noting that it resembles a triangle,

$$A_h = \frac{1}{2}(d_{out} - d_{in})(u_{out} - u_{in}). \tag{6.19}$$

Table 6.8 gives the value of Hamaker's constant, A_H, the area enclosed by the hysteresis loop, A_h, and the ratio A_h/A_H^2. This ratio will stay almost constant for a large range of values of Hamaker's constant, because the Lennard–Jones potential is linear in A_H. Consequently, the value of the parameters associated with the snap-in and snap-out processes, $d_{out} - d_{in}$ and $u_{out} - u_{in}$, will each scale with A_H, so that $A_H \propto A_h^2$.

Table 6.8 Hamaker's constant, A_H, the area enclosed by the hysteresis loop, A_h, and the ratio A_h/A^2.

$$A_H = 0.425 \,\text{aJ}$$
$$A_h = 122.884 \,\text{nm}^2$$
$$A_h/A_H^2 = 680.326 \,\text{m/aJ}$$

6.5 Animation

Figure 6.13 shows a time-elapsed series of cantilever–tip–sample positions generated by using the parameters given in Table 6.1. Here, the sample is moving up and down, and the cantilever responds to the tip–sample force. Note that at stage 1 the cantilever is pushed up by the sample, while it is pulled down by the attractive tip–sample forces at stages 2–7, and at stage 8 the cantilever snaps back to its neutral position. A live animation is presented in the code belonging to this chapter.

▶ **Exercises for Chapter 6**

1. Modify the parameters in Table 6.1 and discuss the resultant snap-in and snap-out points of the various hysteresis loops.
2. Modify the parameters in Table 6.1 and check the relationship between the area enclosed by several hysteresis loops and the product of the tip radius and Hamaker's constant.

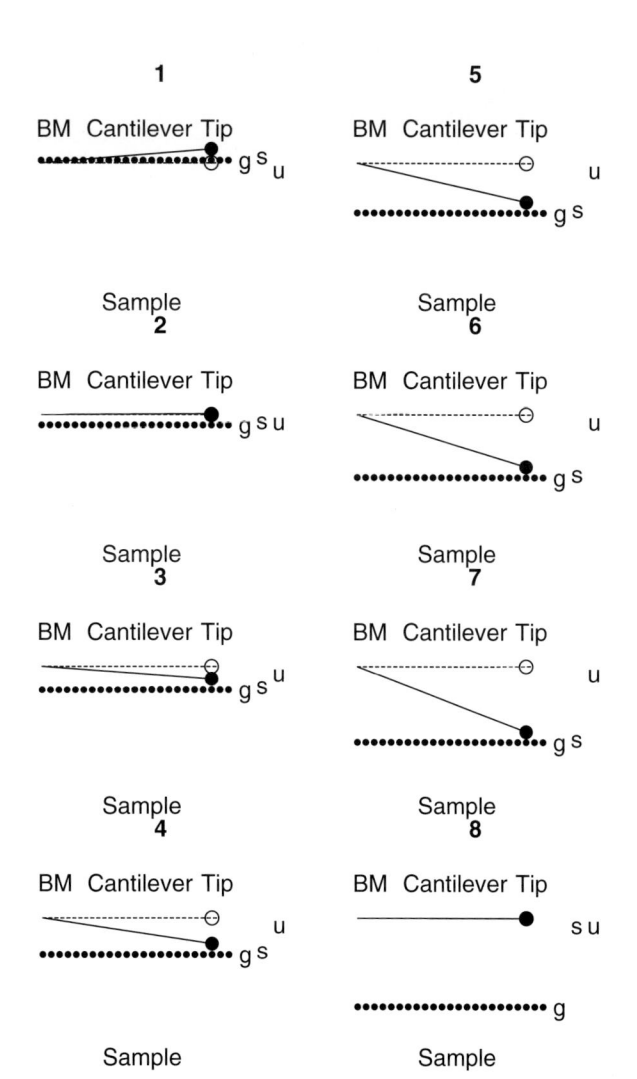

Fig. 6.13 A time-elapsed series of cantilever-tip–sample positions.

References

[1] D. Sarid, *Scanning Force Microscopy with Applications to Electric, Magnetic, and Atomic Forces*, Oxford University Press, New York, 1994.

7
Free Vibrations

Highlights

1. Transient and steady-state solutions to the equation of motion of a free cantilever
2. Phase, amplitude, and energy of the vibration as a function of frequency and damping

Abstract

The vibration of the cantilever of an atomic force microscope, in the absence of a tip–sample force, is modeled as a driven, damped, linear oscillator. For simplicity, the oscillator is modeled as a point mass suspended on a spring mounted on a vibrating bimorph. Analytical and numerical solutions are presented for the response of the cantilever in the vicinity of its resonance frequency. In particular, the amplitude and phase of vibration as a function of frequency and damping are presented together with phase diagrams.

7.1
Introduction

This is the first in a series of three chapters dealing with the vibration of the cantilever of an atomic force microscope. This chapter treats the cantilever as a driven, damped, linear oscillator, neglecting tip–sample forces. The next two chapters extend these discussions to the cases where the tip of the cantilever interacts with the surface of a sample, but does not contact it, and where the tip touches the surface intermittently.

Figure 7.1 is a schematic diagram of the cantilever mounted on a bimorph that vibrates at a given amplitude and frequency, a and f, respectively. The cantilever attached to it will respond by vibrating with an amplitude and phase given by A and ϕ, respectively, that depend on its mass, m, spring constant, k, quality factor associated with its damping, Q, and resonance frequency, f_0. If the amplitude of the vibrating cantilever is small enough, one can approximate its response by using an equivalent point mass suspended by a spring mounted on a vibrating bimorph. Special emphasis is given to the response of the cantilever operating at its resonance frequency and at a frequency that is at the steepest slope of its resonance curve. The equation of motion of this system is solved both analytically and numerically.

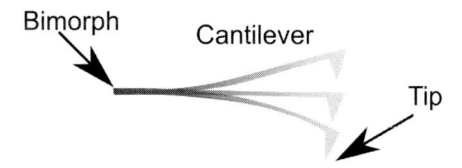

Fig. 7.1 A schematic diagram of a driven cantilever.

The four parameters that dictate the operation of the vibrating cantilever, a, k, f_0, and Q, which will be used in the examples, are given in Table 7.1. Note that the values of Q and f are much smaller than those of typical cantilevers. However, using these values makes it easier to explore the response of the cantilever, without any loss of accuracy. Unless specifically stated, we will use the value of Q given in Table 7.1.

Table 7.1 The four parameters that dictate the operation of the vibrating cantilever, a, k, f_0, and Q.

$a = 1\,\text{nm}$
$k = 1\,\text{N/m}$
$f_0 = 10\,\text{Hz}$
$Q = 5$

7.2 Equation of Motion

We start by deriving the equation of motion of a driven cantilever whose tip is far enough from the surface of a sample such that tip–sample forces can be neglected. Let u denote the vibration of the bimorph driving the cantilever,

$$u(t) = u_0 + a e^{-i\omega t}, \tag{7.1}$$

where u_0 is the bimorph average position, $\omega = 2\pi f$, and t is time. The resultant vibration of the cantilever at its free end will be denoted by

$$z(t) = z_0 + z(t), \tag{7.2}$$

where z_0 is its average position. The equation of motion of a point mass, m, representing the vibrating cantilever, is

$$F_0 = m\frac{\partial^2}{\partial t^2}z(t) + \gamma\frac{\partial}{\partial t}z(t) + k(z(t) - u). \tag{7.3}$$

Here, γ is a damping (loss) coefficient proportional to the velocity of the cantilever, given later in terms of Q, and F_0 is a constant force. Note that in a real experiment, where the cantilever and tip are surrounded by a fluid (gas or liquid), both the effective mass of the cantilever and its damping constant are functions of the tip–sample distance. Here, however, we neglect such an effect. Inserting Eqs. (7.1) and (7.2) into Eq. (7.3) yields

$$F_0 = m\frac{\partial^2}{\partial t^2}z(t) + \gamma\frac{\partial}{\partial t}z(t) + k[z_0 + z(t) - u_0 - ae^{-i\omega t}]. \tag{7.4}$$

For a vanishing F_0, we get

$$z_0 = u_0, \tag{7.5}$$

and Eq. (7.4) reduces to

$$m\frac{\partial^2}{\partial t^2}z(t) + \gamma\frac{\partial}{\partial t}z(t) + kz(t) = ake^{-i\omega t}. \tag{7.6}$$

Recall that the resonance frequency of a lossless cantilever, for which $\gamma = 0$, is

$$\omega_0 = \sqrt{k/m}, \tag{7.7}$$

and that for a lossy cantilever, γ and Q are related by

$$Q = m\omega_0/\gamma. \tag{7.8}$$

Using Eq. (7.7) and Eq. (7.8) in Eq. (7.6) yields the equation of motion

$$\frac{\partial^2}{\partial t^2}z(t) + \frac{\omega_0}{Q}\frac{\partial}{\partial t}z(t) + \omega_0^2 z(t) = a\omega_0^2 e^{-i\omega t}, \tag{7.9}$$

which will be first solved analytically and then numerically by using the parameters given in Table 7.1.

7.3
Analytical Solution

7.3.1
Equation of Motion

We will now solve the equation of motion analytically, find an expression for the frequency-dependent response of the cantilever, and use it to plot a typical case. First, consider the case of a free, lossless cantilever that oscillates harmonically. For a vanishing value of a and an infinite value of Q, Eq. (7.9) reduces to

$$\frac{\partial^2}{\partial t^2}z(t) + \omega_0^2 z(t) = 0. \tag{7.10}$$

The solution for z is readily obtained as

$$z = e^{-i\omega_0 t}. \tag{7.11}$$

Next, consider the case of a free, lossy cantilever having a finite Q. Equation (7.9) will now reduce to

$$\frac{\partial^2}{\partial t^2}z(t) + \frac{\omega_0}{Q}\frac{\partial}{\partial t}z(t) + \omega_0^2 z(t) = 0. \tag{7.12}$$

Assume that the motion of the cantilever, z, can be described by

$$z = e^{pt}. \tag{7.13}$$

Inserting Eq. (7.13) into Eq. (7.12) yields

$$p^2 + p\frac{\omega_0}{Q} + \omega_0^2 = 0, \tag{7.14}$$

and solving for p gives

$$p = -\frac{\omega_0}{2Q} - \frac{\omega_0}{2Q}\sqrt{1 - 4Q^2}. \tag{7.15}$$

For $4Q^2 > 1$, which is always the case, Eq. (7.15) can be written in terms of real and imaginary parts as

$$p = -i\omega_0\sqrt{1 - \frac{1}{4Q^2}} - \frac{\omega_0}{2Q}. \tag{7.16}$$

The imaginary part of p gives the resonance frequency of the freely oscillating, lossy cantilever, ω_1,

$$\omega_1 = \omega_0\sqrt{1 - \frac{1}{4Q^2}}, \tag{7.17}$$

and the real part of p gives the decay of the vibrations, so that the solution to Eq. (7.12) yields

$$z = e^{-i\omega_1 t} e^{-\frac{1}{2Q}\omega_0 t}. \tag{7.18}$$

7.3.2
Steady-State Regime

We will now solve the general equation of motion of the driven, lossy cantilever, Eq. (7.9). Assume now that the amplitude of vibration of the cantilever, z, can be described by

$$z = Ae^{-i\omega t}, \tag{7.19}$$

where A is a complex amplitude. Inserting Eq. (7.19) into Eq. (7.9) and cancelling common factors leads to

$$A\left(1 - x^2 - ix\frac{1}{Q}\right) = a, \qquad (7.20)$$

where $x = \omega/\omega_0$. Solving for A yields

$$A = \frac{aQ}{Q(1-x^2) - ix}, \qquad (7.21)$$

with its absolute value, A_0, given by

$$A_0 = \frac{aQ}{\sqrt{Q^2(1-x^2)^2 + x^2}}. \qquad (7.22)$$

As expected, A_0 is a function of ω, ω_0, and Q, and is proportional to the amplitude of vibration of the bimorph, a.

7.3.3
Bimorph–Cantilever Phase

The phase of the vibration of the cantilever, ϕ, relative to that of the driving bimorph, is obtained from the ratio of the imaginary and real components of A,

$$\phi = \arctan\left[\frac{1}{Q(\frac{1}{x} - x)}\right]. \qquad (7.23)$$

We can also write A_0 in terms of $\sin\phi$, as

$$A_0 = aQ\frac{1}{x}\sin\phi, \qquad (7.24)$$

where

$$\sin\phi = \frac{1}{\sqrt{1 + Q^2\left(x - \frac{1}{x}\right)^2}}. \qquad (7.25)$$

Figure 7.2 shows the bimorph–cantilever phase, ϕ, as a function of the normalized frequency $x = \omega/\omega_0$, for $Q = 1$ (solid line), $Q = 5$ (dashed line), and $Q = 10$ (dotted line). Note that the steepness of the curve reflects the magnitude of Q. Note also that for frequencies well below resonance $\phi \to 0$, at resonance $\phi = \pi/2$, and far above resonance $\phi \to \pi$.

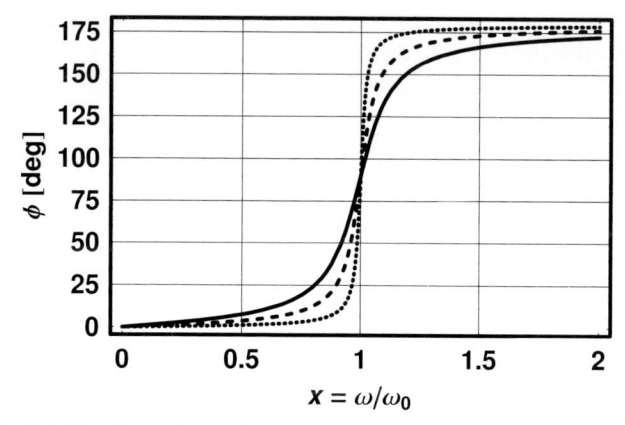

Fig. 7.2 The bimorph–cantilever phase, ϕ, as a function of the normalized frequency $x = \omega/\omega_0$, for $Q = 1$ (solid line), $Q = 5$ (dashed line), and $Q = 10$ (dotted line).

7.3.4
Q-Dependent Resonance Frequency

The finite value of Q modifies the resonance frequency of the driven cantilever from ω_0 to ω_m, given by

$$\omega_m = \sqrt{1 - \frac{1}{2Q^2}}\,\omega_0. \tag{7.26}$$

Table 7.2 gives the resonance frequencies of the free and lossless, free and lossy, and driven and lossy cantilevers, ω_0, ω_1 and ω_m, respectively, together with their respective amplitudes, A_0, for $Q = 5$.

Table 7.2 The resonance frequencies of the free and lossless, free and lossy, and driven and lossy cantilevers, ω_0, ω_1 and ω_m, respectively, together with their respective amplitudes, A_0.

	ω_0	ω_1	ω_m
ω	62.831 9	62.516 9	62.200 4
A_0	5	5.018 86	5.025 19

Figure 7.3 shows the Q-dependent, normalized resonance frequency, ω_m/ω_0, for the range $5 < Q < 15$.

One observes that decreasing Q will lower the resonance frequency, albeit by a rather small value.

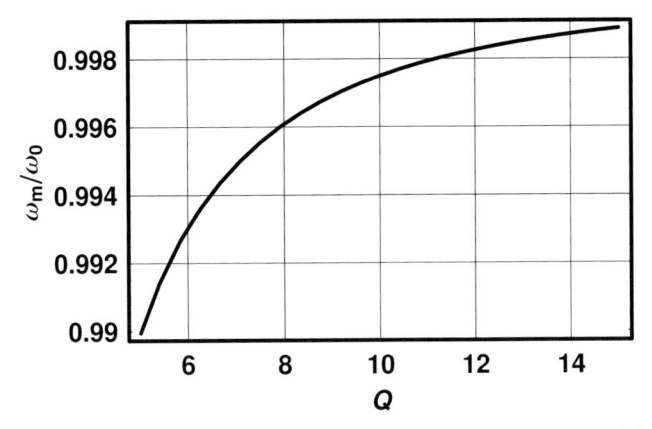

Fig. 7.3 The scaled resonance frequency, ω_m/ω_0, as a function of Q.

7.3.5
Frequency-Dependent Amplitude

The frequency-dependent amplitude of vibration of the cantilever, A_0, as a function of the normalized frequency, ω/ω_0, for $Q = 1$ (solid line), $Q = 5$ (dashed line), and $Q = 10$ (dotted line), is shown in Figure 7.4. Here, the peak of the resonance curve, marked by a vertical line, is somewhat below $\omega = \omega_0$, because Q is finite. Note that the below- and above-resonance regions are dominated by the spring constant of the cantilever and its mass, respectively. On resonance, however, the peak is determined by the quality factor Q.

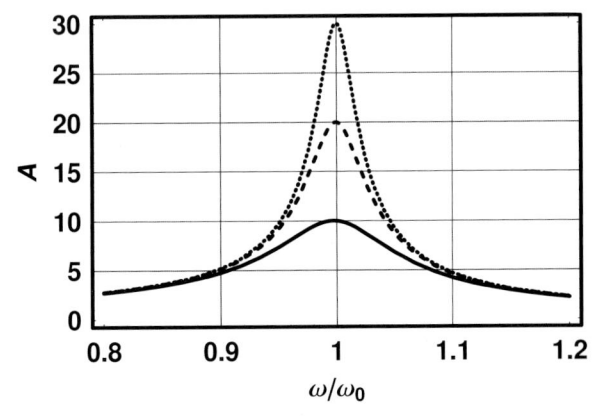

Fig. 7.4 The frequency-dependent amplitude of vibration, A_0, as a function of the scaled resonance frequency, ω/ω_0, for $Q = 1$ (solid line), $Q = 5$ (dashed line), and $Q = 10$ (dotted line).

7.3.6
Frequency at the Steepest Slope

Operating at one of the two frequencies defined by the steepest slope of the resonance curve is of experimental importance. The reason for it is that at these frequencies, the amplitude of vibration is most sensitive to changes in the tip–sample force, as will discussed in the next chapter. An approximate value for these frequencies, widely used in the literature, is given by $\omega_{m_1} = \omega_0(1 - 0.35/Q)$ and $\omega_{m_2} = \omega_0(1 + 0.35/Q)$. Table 7.3 gives the ω_{m_1}/ω_0 and ω_{m_2}/ω_0 together with their exact value.

Table 7.3 The normalized frequencies ω_{m_1}/ω_0 and ω_{m_2}/ω_0 together with their exact value.

	ω_{m_1}/ω_0	ω_{m_2}/ω_0
Exact	57.8417	66.7227
Approximate	58.4336	67.2301

Note that the exact and approximate solutions do not agree for small values of Q, such as $Q = 5$. However, for larger values of Q, for example, $Q = 50$, the agreement is good.

7.3.7
Average Power

The average power delivered to the vibrating cantilever by the externally driving force, P_{av}, is obtained by time averaging the velocity and force product. Here, the average of a cycle yields

$$P_{av} = \frac{1}{2} a k A_0 \omega \sin \phi, \tag{7.27}$$

where $\sin \phi$ accounts for the angle between the driving force and the response of the cantilever. Figure 7.5 shows the average power, P_{av}, as a function of the normalized frequency ω/ω_0 for $Q = 10$ (solid line), $Q = 20$ (dashed line), and $Q = 30$ (dotted line).

7.4
Numerical Solutions

7.4.1
Equation of Motion

We are now in a position to code the general solution of the equation of motion of the cantilever in the presence of the tip–sample force. The equation of

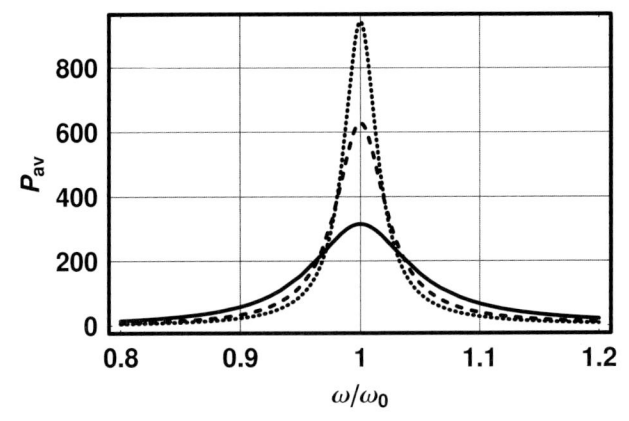

Fig. 7.5 The frequency-dependent average power, P_{av}, as a function of the scaled frequency, ω/ω_0, for $Q = 10$ (solid line), $Q = 20$ (dashed line), and $Q = 30$ (dotted line).

motion can be written as two first-order differential equations, one given by

$$z_2'(t) = -\frac{\omega_0}{Q} z_2(t) - \omega_0^2 [z_1(t) - a\sin(\omega t)], \qquad (7.28)$$

where $z_1(t) = z(t)$, and the other given by $z_2(t) = z_1'(t)$. By solving these two coupled equations as interpolation functions, we obtain the numerical solution to the response of the cantilever to the driving bimorph.

7.4.2
Transient Regime

Figure 7.6 shows the transient regime of the vibrations of the cantilever in the first $2Q$ cycles, for $\omega = \omega_0$, together with their envelope (solid line) as given by Eq. (7.18).

It is instructive to compare the vibrations of the cantilever for $\omega = \omega_0$ with the $\pi/2$ phase-shifted vibrations of the bimorph, shown in Figure 7.7, as a function of time, t. Note that the amplitude of vibrations of the bimorph has been multiplied by aQ.

Figure 7.7 confirms that the vibration of the cantilever on resonance indeed lags behind that of the bimorph by $\pi/2$.

7.4.3
Bimorph–Cantilever Phase Diagram

Figure 7.8 shows the bimorph–cantilever phase diagram for $2Q$ cycles, obtained as a parametric plot with the common variable t. The phase of the vi-

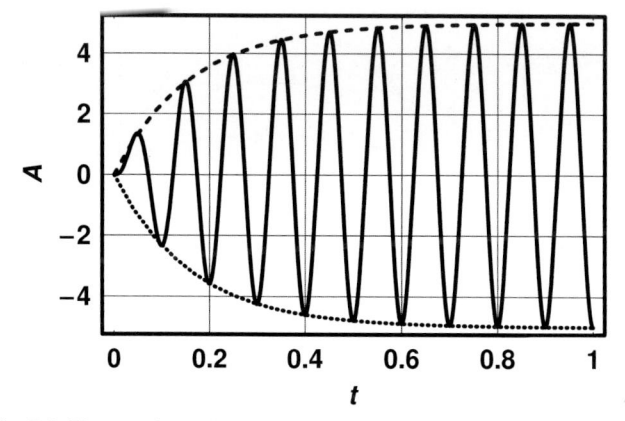

Fig. 7.6 The transient vibrations of the cantilever and their envelope (solid and dashed lines).

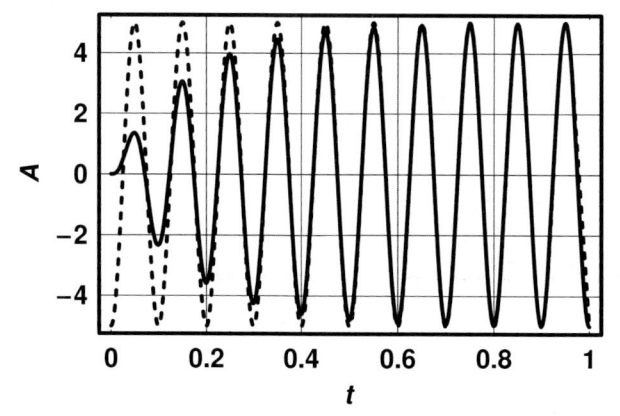

Fig. 7.7 The shifted, transient vibration of the cantilever and the vibration of the bimorph (dashed line) for $\omega = \omega_0$.

brations of the cantilever, relative to those of the bimorph driving it, evolves to an almost steady state after only $2Q$ cycles. The perfectly circular contour of the phase plot after $2Q$ cycles implies that the vibrations is well behaved.

7.4.4
Displacement–Velocity Phase Diagram

It is instructive to observe the cantilever displacement–velocity phase diagram, where $z(t)$ and $\partial z(t)/\partial t$ are parametrically plotted using t as the common variable. The plot shown in Figure 7.9 demonstrates again that the vibrations are well behaved.

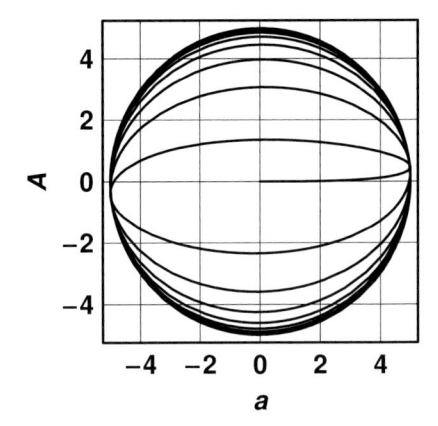

Fig. 7.8 The bimorph–cantilever phase diagram.

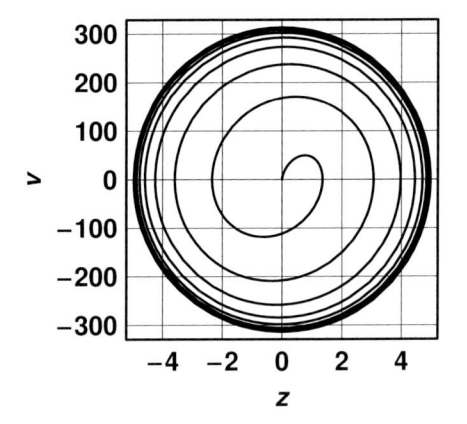

Fig. 7.9 The displacement–velocity phase diagram.

▶ **Exercises for Chapter 7**

1. Calculate and plot the amplitude and phase of vibration of the cantilever for a range of parameters a, k, f_0, and Q.
2. Repeat Problem 1 but now use the frequencies at the steepest slope of the resonance curve.

References

1. K. R. Symon, *Mechanics*, 3rd Edition, Addison-Wesley, London, 1971.
2. Y. Martin, C. C. Williams, and H. K. Wickramasinghe, "Atomic force microscope-force mapping and profiling on a sub 100-Å scale," J. Appl. Phys. **61**, 4723 (1987).
3. J. P. Cleveland, B. Anczykowski, A. E. Schmid, and V. B. Ealings, "Energy dissipation in tapping-mode atomic force microscopy," Appl. Phys. Lett. **72**, 2613 (1998).

8
Noncontact Mode

Highlights

1. Transient and steady-state solutions to the equation of motion
2. Phase and amplitude of vibration as a function of frequency

Abstract

The vibration of the cantilever of an atomic force microscope, in the presence of a tip–sample force, is modeled as a driven, damped, nonlinear oscillator. The dependence of the resonance frequency, amplitude, and phase of vibration of the cantilever on the tip–sample force is solved numerically and analytically and a comparison between the two is presented using typical parameters. It is shown that an approximate analytical solution, involving the tip–sample force derivative, can be used to model effects of tip–sample electric, magnetic, and atomic forces.

8.1
Introduction

The previous chapter treated the basic properties of a free cantilever, excited by a vibrating bimorph and placed far away from a sample. Here we treat the case where the tip of the cantilever is brought into close proximity to the surface of a sample, yet does not make contact with it. The tip–sample force, modeled by a Lennard–Jones potential, will therefore always stay in the attractive regime. An AFM whose tip position is limited to this regime is considered to operate in the noncontact mode. A schematic diagram of the tip–sample interaction given in Figure 8.1 shows a bimorph, cantilever, tip, and sample, all vibrating at a frequency f. The bimorph vibrates with an amplitude a, and the cantilever with an amplitude A and phase ϕ relative to that of the bimorph. The vibration of the cantilever depends on its spring constant k, resonance frequency f_0, quality factor Q, tip–sample force F, and tip–sample average spacing s_{set}. The equation of motion of the cantilever in the presence of the tip–sample force, is presented and solved analytically and numerically using the parameters given in Table 8.1. The solution to the equation of motion of the driven cantilever, in the transient- and steady-state regimes, is discussed in terms of the amplitude and phase of vibration. Expanding the solution in powers of the tip displacement, and using the first-order term in the analytical solution, highlights the importance of the role of the first derivative of the force, F', in modifying the effective spring constant of the cantilever, $k_1 = k - F'$. For an attractive Lennard–Jones force, for which F' is positive, $k_1 < k$, resulting in a decrease of the resonance frequency of the cantilever. The case where the repulsive regime of the force has to be taken into account will be dealt with in the next chapter.

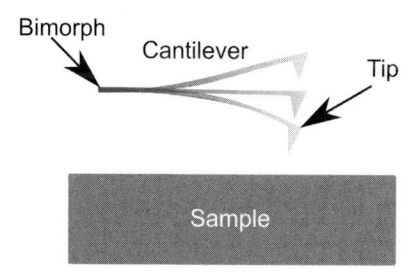

Fig. 8.1 A schematic diagram of the vibrating cantilever, its tip, and a sample.

Table 8.1 The amplitude of vibration of the cantilever, a, its spring constant, k, quality factor, Q, resonance frequency, f_0, and average tip–sample spacing, s_{set}.

$$a = 0.1\,\text{nm}$$
$$k = 10\,\text{N/m}$$
$$Q = 20$$
$$f_0 = 100\,\text{kHz}$$
$$f = 100\,\text{kHz}$$
$$s_{set} = 3\,\text{nm}$$

The tip–sample and the Lennard–Jones potential are characterized by the radius of the tip, R, a typical interatomic spacing, σ, and Hamaker's constant, A_H, given in Table 8.2.

Table 8.2 The the radius of the tip, R, a typical interatomic spacing, σ, and Hamaker's constant, A_H.

$$R = 10\,\text{nm}$$
$$\sigma = 0.34\,\text{nm}$$
$$A_H = 1\,\text{aJ}$$

8.2
Tip–Sample Interaction

8.2.1
Lennard–Jones Potential

This section is a summary of a more comprehensive discussion of the tip–sample Lennard–Jones potential presented in Chapter 6. The purpose of including it here is to make this chapter self-contained. The Lennard–Jones potential, W, is given by

$$W = \frac{A_H R}{6\sigma} \left[\frac{1}{210} \left(\frac{\sigma}{s}\right)^7 - \frac{\sigma}{s} \right], \tag{8.1}$$

The first term on the right-hand side is the magnitude of the potential in terms of A and R, and the second one describes the shape of the potential. The Lennard–Jones force, F, is obtained from the potential by taking the negative of its derivative with respect to s

$$F = \frac{A_H R}{6\sigma^2} \left(\frac{\sigma^2}{s^2} - \frac{\sigma^8}{30 s^8} \right), \tag{8.2}$$

which vanishes at s_{F_0}

$$s_{F_0} = \frac{1}{30^{1/6}} \sigma. \tag{8.3}$$

The Lennard–Jones force as a function of the tip–sample spacing, s, is shown in Figure 8.2.

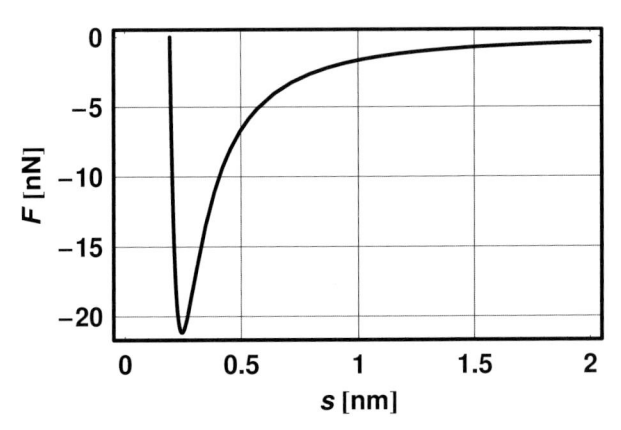

Fig. 8.2 The Lennard–Jones force, F, as a function of the tip–sample spacing, s.

The tip–sample force in the region shown in Figure 8.2 is attractive, and by convention negative. It has been shown in Chapter 6 that on approach, the tip will snap into the sample if the cantilever spring constant is larger than the tip–sample force derivative, F', given by

$$F' = \frac{A_H R}{3\sigma^3}\left(\frac{\sigma^3}{s^3} - \frac{2}{15}\frac{\sigma^9}{s^9}\right). \tag{8.4}$$

The force derivative vanishes at $s_{F_0'}$,

$$s_{F_0'} = \left(\frac{2}{15}\right)^{1/6}\sigma, \tag{8.5}$$

and obtains its maximum value at $s_{F_{max}'}$,

$$s_{F_{max}'} = \left(\frac{2}{5}\right)^{1/6}\sigma. \tag{8.6}$$

Its value at this point, F_{max}', is

$$F_{max}' = \frac{\sqrt{10}}{9}\frac{AR}{\sigma^3}. \tag{8.7}$$

The parameters given in Table 8.1 and Table 8.2 generate a force derivative whose maximum is larger than the spring constant, as shown in Figure 8.3. On approach, therefore, the tip will snap into the sample at a spacing for which $F' = k$, shown as a horizontal line in the figure.

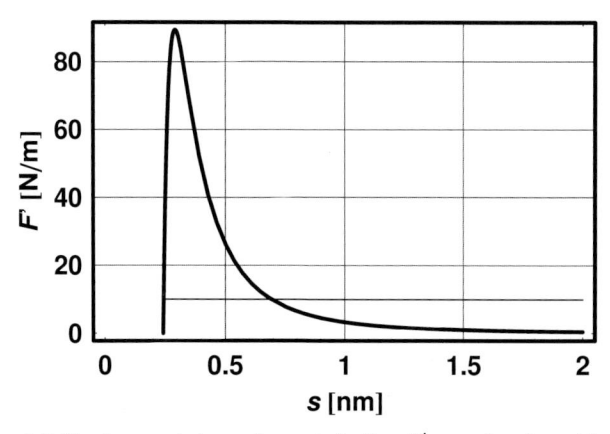

Fig. 8.3 The Lennard–Jones force derivative, F', as a function of the tip–sample spacing, s, and the spring constant of the cantilever, k, shown by a horizontal line.

The results of these calculations yield the values of s_{F_0}, $s_{F_0'}$, F_{max}', and $s_{F_{max}'}$, given in Table 8.3.

Table 8.3 The values of s_{F_0}, $s_{F_0'}$, F_{max}' and $s_{F_{max}'}$ using Table 8.1 and Table 8.2.

s_{F_0}	= 0.192 882 nm
$s_{F_0'}$	= 0.243 016 nm
F_{max}'	= 0.243 016 nm
$s_{F_{max}'}$	= 0.291 847 nm

8.2.2
The Equation of Motion

Let us start by characterizing the tip–sample force, as derived from a Lennard–Jones potential. We first formulate the equation of motion of a cantilever whose tip experiences this force and solve the equation numerically (no exact analytical solution exists). Finally, we analyze the resulting features characterizing the vibration of the cantilever. It was shown in the previous chapter that the cantilever reaches almost a steady state in its vibration after about $2Q$ cycles. We will therefore stop the numerical calculations at this value. Table 8.4 gives the number of cycles, their period, and the duration of the calculation.

Table 8.4 The number of cycles, their period, and the duration of the calculation.

Cycles	= 40
Period	= 10 µs
Duration	= 400 µs

We are now in a position to code the general solution of the equation of motion of the cantilever in the presence of the tip–sample force. The equation of motion can be written as two first-order differential equations, one given by

$$z_2'(t) = -\frac{\omega_0}{Q} z_2(t) - \omega_0^2 [z_1(t) - a\sin(\omega t)], \quad (8.8)$$

where $z_1(t) = z(t)$, and the other given by $z_2(t) = z_1'(t)$. By solving these two coupled equations as interpolation functions, we obtain the numerical solution to the response of the cantilever to the driving bimorph.

8.2.3
Numerical Solution of the Equation of Motion

Before analyzing the numerical solution, which is given as an interpolation function, it is useful to recall the analytical solution to the vibration of a driven cantilever, z_{free}, with no tip–sample forces,

$$z_{free}(t) = z_{set} + aQ\left(1 - e^{-\frac{\omega_0}{2Q}t}\right)\sin(\omega_0 t - \pi/2). \tag{8.9}$$

Note that here the phase of z_{free} is shifted by $-\pi/2$. The interesting parts of the numerical and analytical solutions appear in the transient- and steady-state regimes, which will be described in the following sections. The numerical solutions, z (solid line), and the shifted analytical solution, z_{free} (dashed line), for the transient vibration regime, are shown in Figure 8.4 for a tip–sample spacing $s = s_{set}$. It is observed that these two solutions overlap each other for the first few cycles.

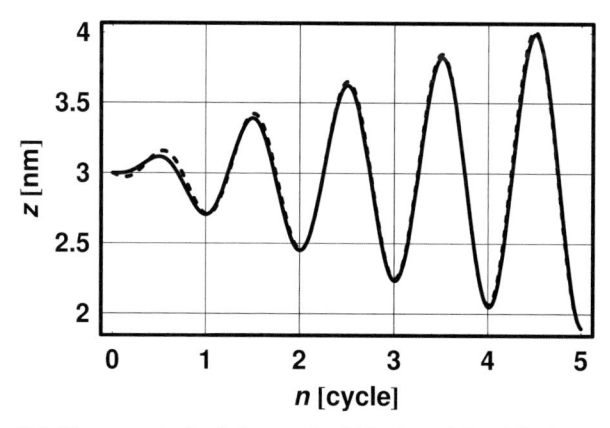

Fig. 8.4 The numerical solutions, z (solid line), and the shifted analytical solution, z_{free} (dashed line), for the first five cycles.

Let us now compare z and z_{free} for the steady-state regime in the last two cycles. The comparison of z (solid line) and z_{free} (dashed lines) in the steady-state regime is shown in Figure 8.5. Here we observe that the presence of the tip–sample force produces a shift in the phase of the z relative to that of z_{free}, a topic that will be investigated in the following sections.

To further explore the phase of vibration of the cantilever, we produce a parametric plot of z against z_{free}, shown in Figure 8.6. If these two functions were in phase, the curve would produce a straight diagonal line. However, the presence of a finite tip–sample force gives rise to a phase shift between these two vibrations, as evidenced here by the elliptical curve.

The phase diagram of z against $v = z'(t)$ is shown in Figure 8.7 for the steady-state regime. Here, the parametric plot yields an almost perfect circle, indicating the "good behavior" of the vibration of the cantilever, namely, that it follows an almost harmonic motion.

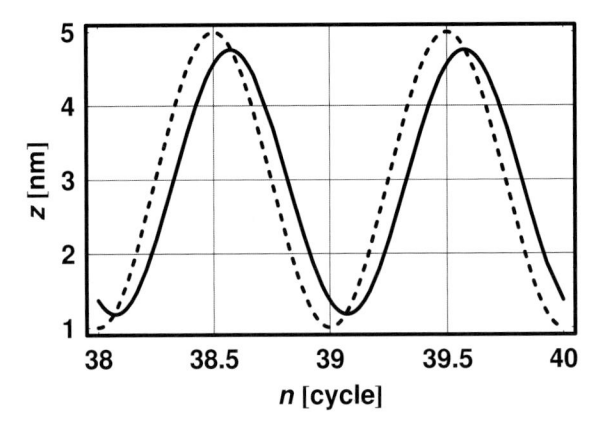

Fig. 8.5 A comparison of z and z_{free} for the steady–state regime in the last two cycles.

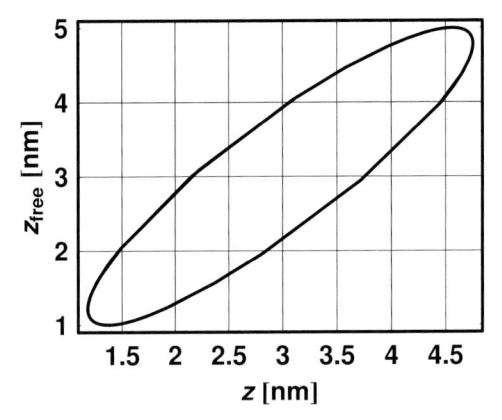

Fig. 8.6 The phase diagram of z against z_{free} in the last two cycles.

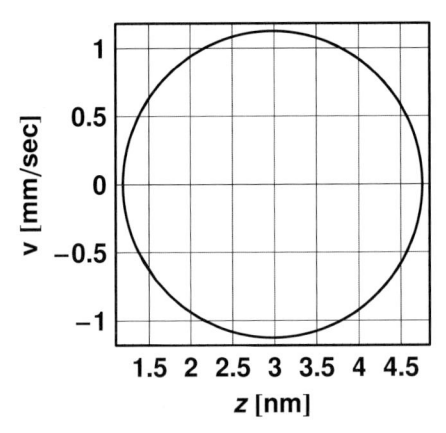

Fig. 8.7 The displacement–velocity phase diagram in the last two cycles.

8.2.4
Approximate Analytical Solution of the Equation of Motion

Assume that the tip of the cantilever vibrates far enough from the sample such that the tip–sample force it experiences can be considered as only a small perturbation to the force-free case. One can expand the tip–sample force in powers of the tip displacement, and use the first-order term in the analytical solution to the motion of the cantilever. In this section, we solve the equation of motion analytically using a first-order perturbation, and compare the solution to the numerically exact one. Let the bimorph driving the cantilever vibrate according to

$$u = u_0 + a e^{i\omega t}, \tag{8.10}$$

where u_0 is its average position. Let the vibration of the tip be

$$z = z_0 + z(t), \tag{8.11}$$

where z_0 is the average position of the tip, and expand the nonuniform tip–sample force, F, around z_0,

$$F = \sum_{n=0}^{\infty} F^{(n)} (z - z_0)^n. \tag{8.12}$$

The assumption here is that the variation of the force along the excursion of the vibrating tip is small enough to justify a first-order perturbation solution. To first order, one can rewrite Eq. (8.12) as

$$F = F_0 + F' z(t), \tag{8.13}$$

where both F_0 and F' are evaluated at z_0. The equation of motion, therefore, becomes

$$m z''(t) + \frac{m \omega_0}{Q} z'(t) + k(z - u) = F_0 + F' z(t). \tag{8.14}$$

Inserting Eqs. (8.10), (8.11), and (8.13) into Eq. (8.14) yields

$$m z''(t) + \frac{m \omega_0}{Q} z'(t) + k \left[z_0 + z(t) - u_0 - a e^{i\omega t} \right] = F_0 + F' z(t). \tag{8.15}$$

At equilibrium,

$$F_0 = k(z_0 - u_0), \tag{8.16}$$

and Eq. (8.15) reduces to

$$m z''(t) + \frac{m \omega_0}{Q} z'(t) + k_1 z(t) = a k e^{i\omega t}, \tag{8.17}$$

where

$$k_1(z) = k - F'(z). \tag{8.18}$$

The resonance frequency of a lossless cantilever, ω_1, for which $Q \to \infty$, determined by k_1 rather than by k, is given by

$$\omega_1 = \sqrt{k_1/m}. \qquad (8.19)$$

Inserting Eq. (8.19) into Eq. (8.17) leads to

$$z''(t) + \frac{\omega_0}{Q} z'(t) + \omega_1^2 z(t) = a\omega_0^2 e^{i\omega t}, \qquad (8.20)$$

where both ω_0 and ω_1 are included. We will now solve Eq. (8.20) analytically for the steady-state regime and explore its properties. Previously, the force derivative, F', was a function of z. We now wish to use it as a variable and analyze the effective spring constant k_1 in conjunction with the resonance frequency ω_1,

$$k_1(F') = k - F'. \qquad (8.21)$$

Assume now that

$$z(t) = A e^{i\omega t}, \qquad (8.22)$$

so that Eq. (8.20) becomes

$$-A\omega^2 + iA\frac{\omega\omega_0}{Q} + A\frac{k_1}{k}\omega_0^2 = a\omega_0^2. \qquad (8.23)$$

The steady state, complex solution of the equation of motion, A, is given by

$$A = \frac{akQ\omega_0^2}{-kQ\omega^2 + ik\omega\omega_0 - F'Q\omega_0^2 + kQ\omega_0^2}, \qquad (8.24)$$

which yields the amplitude of vibration of the cantilever, A_0,

$$A_0 = \frac{akQ\omega_0^2}{\sqrt{k^2\omega^2\omega_0^2 + (kQ\omega^2 + F'Q\omega_0^2 - kQ\omega_0^2)^2}}. \qquad (8.25)$$

Using Eq. (8.19) and Eq. (8.21) in Eq. (8.25) leads to

$$A_0 = \frac{aQ}{\sqrt{\frac{\omega^2}{\omega_0^2} + Q^2\left(\frac{\omega^2}{\omega_0^2} - \frac{k_1}{k}\right)^2}}. \qquad (8.26)$$

The phase between the vibration of the bimorph and the cantilever, $\Delta\phi$, is obtained from the ratio of the imaginary and real components of A,

$$\Delta\phi = \arctan\left(\frac{k\omega\omega_0}{kQ(\omega^2 - \omega_0^2) + F'Q\omega_0^2}\right), \qquad (8.27)$$

which, on resonance, becomes

$$\Delta\phi_1 = \arctan\left(\frac{k}{QF'}\right). \tag{8.28}$$

Recall now the frequencies for which the resonance curve is the steepest, namely, at ω_{m_1} and ω_{m_2}, given in Chapter 7 by

$$\omega_{m_1} = \omega_0(1 - 0.35/Q) \tag{8.29}$$

and

$$\omega_{m_2} = \omega_0(1 + 0.35/Q). \tag{8.30}$$

Figure 8.8 shows the phase of vibration of the cantilever, $\Delta\phi$, as a function of the tip–sample spacing, s, for $\omega = \omega_0$ (solid line), $\omega = \omega_{m_1}$ (dashed line), and $\omega = \omega_{m_2}$ (dotted line). For a large spacing, and for $\omega = \omega_0$, we find that $\Delta\phi \to 90°$, as expected. Also, the phases for $\omega = \omega_{m_1}$ and for $\omega = \omega_{m_2}$ are larger and smaller than the phase for $\omega = \omega_0$, respectively.

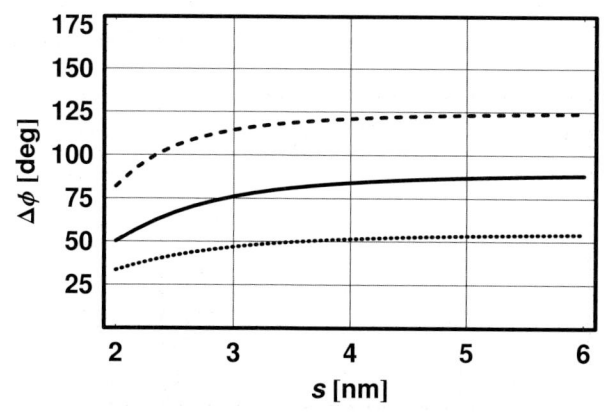

Fig. 8.8 The phase of vibration of the cantilever, $\Delta\phi$, as a function of the tip–sample spacing, s, for $\omega = \omega_0$ (solid line), $\omega = \omega_{m_1}$ (dashed line), and $\omega = \omega_{m_2}$ (dotted line).

Figure 8.9 shows the normalized amplitude of vibration of the cantilever, $A/(aQ)$, as a function of the force derivative, F', for $\omega = \omega_0$ (solid line), $\omega = \omega_{m_1}$ (dashed line), and $\omega = \omega_{m_2}$ (dotted line), in the range $-0.1k < F' < 0.1k$. It should be noted that the effect of the force derivative, during one cycle of vibration, is larger when the tip is at its smallest distance to the sample than when it is at its largest distance. It is therefore not clear exactly at what value of s should we evaluate the force derivative. Nevertheless, we can conclude that using the approximate analytical solution, where k is replaced by $k - F_1$, gives a reasonable approximation of the effect that the tip–sample force has on the vibration of the cantilever. One can therefore employ this approximation to model the response of the cantilever to atomic, electric, and magnetic tip–sample forces, by using their derivatives with respect to the tip–sample spacing.

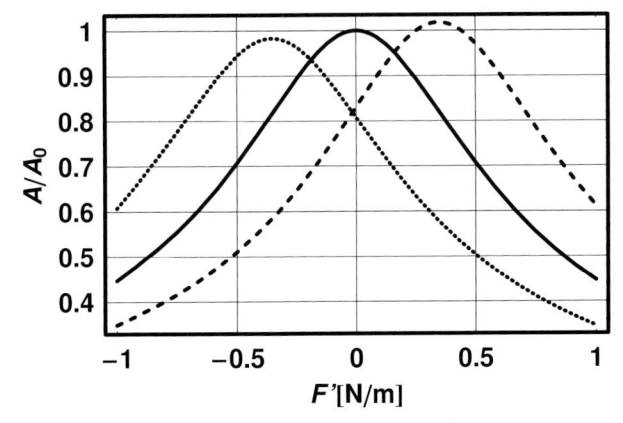

Fig. 8.9 The normalized amplitude of vibration of the cantilever, $A/(aQ)$, as a function of the force derivative, F', for $\omega = \omega_0$ (solid line), $\omega = \omega_{m_1}$ (dashed line), and $\omega = \omega_{m_2}$ (dotted line).

▶ **Exercises for Chapter 8**

1. Calculate and plot the amplitude and phase of vibration of the cantilever for a range of parameters a, k, f, Q, f, and F'.
2. Calculate and plot the case where the sample is vibrated by a bimorph, causing the cantilever to vibrate via the tip–sample force derivative.

References

1. Y. Martin, C. C. Williams, and H. K. Wickramasinghe, "Atomic force microscope force mapping and profiling on a sub 100-Å scale," J. Appl. Phys. **61**, 4723 (1987).
2. D. Sarid, *Scanning Force Microscopy with Applications to Electric, Magnetic and Atomic Forces*, Oxford University Press, New York, 1994.
3. S. N. Magonov, V. Elings, and M.-H. Whangbo, "Phase imaging and stiffness in tapping mode atomic force microscopy," Surf. Sci. Lett. **375**, L385 (1997).

9
Tapping Mode

■ **Highlights**

1. Transient and steady-state solutions to the general equation of motion
2. Tapping amplitude and phase relative to those of a free cantilever
3. Tip–sample indentation force, and pressure

Abstract

The operation of a tapping mode atomic force microscope is analyzed numerically in terms of an attractive tip–sample Lennard–Jones potential and a repulsive indentation force. The results yield the amplitude and phase of the tapping cantilever, and the tip–sample indentation depth, force, and pressure.

9.1
Introduction

This chapter treats the case in which a cantilever is mounted on a bimorph and is vibrated close to its resonance frequency, inducing it to tap on the surface of a sample. While the previous chapter treated only an attractive tip–sample force, here we introduce also a tip–sample repulsive force that indents the sample. A schematic of the tip–sample interaction in the tapping mode is presented in Figure 9.1; it shows a bimorph vibrating at a frequency f and amplitude a, and a cantilever vibrating with an amplitude A and phase ϕ. The amplitude and phase of vibration of the cantilever in this case depend on its spring constant k, resonance frequency f_0, quality factor Q, and tip–sample force F, as well as on the equilibrium position of the cantilever above the sample, s_{set}, which is controlled by the operator of the atomic force microscope. The indentation force, which is repulsive, limits the amplitude of vibration of the cantilever and modifies its phase of vibration. This mode of operation, dubbed the tapping mode, is modeled as a driven, nonlinear, damped oscillator, for which no analytical solution exists. As a first approximation, however, one can use a model of a nearly harmonic grazing-impact oscillator whose analytic solution is well known. For a general solution, however, one has to solve the equation of motion numerically, obtaining vibration characteristics that depend strongly on the set of parameters that define the system. In this chapter, we model the tip and sample as a sphere and a plane, respectively, that interact via an attractive Lennard–Jones force and a repulsive indentation force. The equation of motion is defined and a numerical solution given for the transient and steady-state regions obtained. The dependence of the amplitude and phase of vibration of the cantilever on the tip–sample distance and sample softness are derived and discussed. A summary of the results of the numerical calculations uses the parameters of the cantilever and tip, given in Table 9.1, and the parameters of the sample and the electronics of the AFM, given in Table 9.2. It is important to note that on resonance, the tip will tap the surface only if $s_{set} < aQ$, while for larger values of s_{set}, the operation will be in the noncontact mode. The tip–sample interaction is characterized by the radius of the tip, R, Young's modulus and Poisson's ratios of the tip and sample, E_i and v_i, respectively, a typical interatomic distance, σ, and Hamaker's constant, A_H.

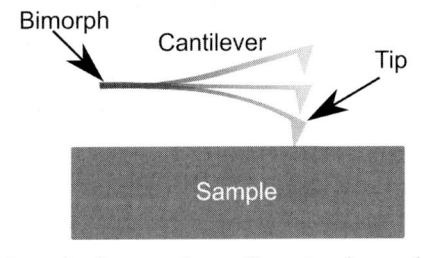

Fig. 9.1 A schematic diagram of a cantilever tapping on the surface of a sample.

Table 9.1 The parameters of the cantilever and the tip.

$k = 20\,\text{N/m}$
$Q = 100\,\text{Hz}$
$R = 10\,\text{nm}$

Note that Young's moduli of the sample and tip are denoted by the subscripts 1 and 2, respectively, given by E_{Pol} and v_{Pol}, and E_{Si} and v_{Si}, respectively. One can choose the operating frequency to be, for example, at resonance, f_0, or at the lower or higher side of the steepest slope of the resonance curve, f_- and f_+, respectively. These three frequencies, given in Table 9.3, were calculated using the parameters given in Table 9.1. For the examples in this chapter, we choose to operate at the lower side of the steepest slope of the resonance curve.

Table 9.2 The parameters of the sample and the electronics of the AFM.

$$A_H = 0.1\,\text{aJ}$$
$$\sigma = 0.34\,\text{nm}$$
$$E_{Si} = 179\,\text{GN/m}^2$$
$$\nu_{Si} = 0.28$$
$$E_{Pol} = 1\,\text{GN/m}^2$$
$$\nu_{Pol} = 0.3$$
$$a = 1\,\text{nm}$$
$$s_{set} = 75\,\text{nm}$$

Table 9.3 The operating frequency at resonance, f_0, and at the lower and higher side of the steepest slope of the resonance curve, f_- and f_+, respectively.

$$f_- = 99.65\,\text{kHz}$$
$$f_0 = 100\,\text{kHz}$$
$$f_+ = 100.35\,\text{kHz}$$

9.2
Lennard–Jones Potential

This section is a summary of a more comprehensive discussion of the tip–sample Lennard–Jones potential presented in Chapter 6. The purpose of including it here is to make this chapter self-contained. The Lennard–Jones potential, W, is given by

$$W = \frac{A_H R}{6\sigma}\left[\frac{1}{210}\left(\frac{\sigma}{s}\right)^7 - \frac{\sigma}{s}\right]. \tag{9.1}$$

The first term on the right-hand side is the magnitude of the potential in terms of A_H and R, and the second one describes the shape of the potential. The Lennard–Jones force, F, obtained from the potential by taking the negative of its derivative with respect to s,

$$F = \frac{A_H R}{6\sigma}\left(\frac{\sigma}{s^2} - \frac{\sigma^7}{30 s^8}\right), \tag{9.2}$$

vanishes at s_{F_0}, which is given by

$$s_{F_0} = \frac{1}{30^{1/6}}\sigma. \tag{9.3}$$

The tip–sample force in the region shown in Figure 9.2 is attractive, and, by convention, negative. It has been shown in Chapter 6 that on approach, the tip will snap into the sample if the cantilever spring constant is larger than the tip–sample force derivative, F',

$$F' = \frac{A_H R}{3\sigma^3}\left(\frac{\sigma^3}{s^3} - \frac{2}{15}\frac{\sigma^9}{s^9}\right). \tag{9.4}$$

The force derivative vanishes at $s_{F'_0}$, which is given by

$$s_{F'_0} = \left(\frac{2}{15}\right)^{1/6} \sigma, \tag{9.5}$$

and obtains its maximum value at $s_{F'_{max}}$, which is given by

$$s_{F'_{max}} = \left(\frac{2}{5}\right)^{1/6} \sigma. \tag{9.6}$$

The value of the force derivative at this point, F'_{max}, is

$$F'_{max} = \frac{\sqrt{10}}{9}\frac{A_H R}{\sigma^3}. \tag{9.7}$$

The parameters given in Table 9.1 and Table 9.2 generate a force derivative whose maximum is larger than the spring constant. On approach, therefore, the tip will snap into the sample at a distance for which $F' = k$. The results of these calculations yield the values of s_{F_0}, $s_{F'_0}$, F'_{max}, and $s_{F'_{max}}$, given in Table 9.4.

Table 9.4 The value of s_{F_0}, $s_{F'_0}$, F'_{max}, and $s_{F'_{max}}$.

s_{F_0}	$= 0.192\,882\,\text{nm}$
$s_{F'_0}$	$= 0.243\,016\,\text{nm}$
F'_{max}	$= 0.243\,016\,\text{nN}$
$s_{F'_{max}}$	$= 0.291\,847\,\text{nm}$

9.3
Indentation Repulsive Force

The elastic moduli of the tip and the sample, κ_i, given by

$$\kappa_i = \frac{1 - v_i^2}{\pi E_i}, \tag{9.8}$$

generate the tip–sample effective elastic modulus, κ_{eff},

$$\kappa_{eff} = \kappa_1 + \kappa_2. \tag{9.9}$$

Instead of κ_{eff}, it will be convenient to use a stiffness parameter, k_s, given by

$$k_s = \frac{4}{3\pi\kappa_{eff}}. \qquad (9.10)$$

9.4 Total Tip–Sample Force

We can now combine the Lennard–Jones force for $s > s_{F_0}$, Eq. (9.2), with the indentation force, F_{in},

$$F_{in} = k_s \sqrt{R}(s_{F_0} - s)^{3/2}, \qquad (9.11)$$

for $s < s_{F_0}$, and get the total force, F_{total}, shown in Figure 9.2. The parameters used in this figure and subsequent ones are for a silicon tip and a typical polymer sample. To demonstrate the effect of tip–sample stiffness, we use in the figure the three values, k_s, $5k_s$, and $10k_s$, showing their respective ascending slopes. Note that the derivative of the force at $s = s_{F_0}$ is discontinuous, since at this point the tip hits the sample and starts indenting it.

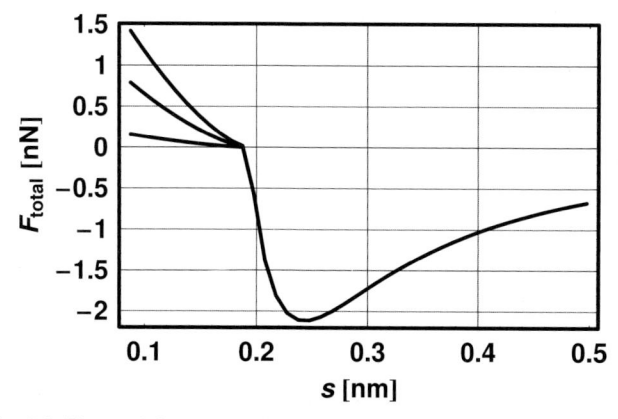

Fig. 9.2 The total tip–sample force, F_{total}, as a function of the distance, s, for k_s, $4k_s$, and $10k_s$, in ascending slope. In this example, the tip is made of silicon and the sample is a typical polymer.

9.5 General Solution

The equation of motion of the bimorph-driven cantilever in the presence of the total force is given by

$$\frac{\partial^2}{\partial t^2}z(t) + \frac{\omega_0}{Q}\frac{\partial}{\partial t}z(t) + \omega_0^2[z(t) - s_{set} - a\sin(\omega t)] = \frac{\omega_0^2}{k}F_{total}[z(t)], \qquad (9.12)$$

with

$$\omega_0 = 2\pi f_0. \tag{9.13}$$

The general solution of the equation of motion can be derived by solving two first-order, coupled differential equations, one given by

$$z_2'(t) = -\frac{\omega_0}{Q} z_2(t) - \omega_0^2 [z_1(t) - a\sin(\omega t)], \tag{9.14}$$

where $z_1(t) = z(t)$, and the other given by $z_2(t) = z_1'(t)$. It was shown before that the vibration of the cantilever approaches a steady state in about $2Q$ cycles, at which stage we will therefore terminate the numerical solution of Eq. (9.12). Before analyzing the numerical solution, which is given as an interpolation function, it is useful to recall the analytical solution to the motion of a driven cantilever in the absence of a tip–sample force, z_{free},

$$z_{\text{free}} = s_{\text{set}} + \frac{aQ}{\sqrt{x^2 + Q^2(1-x^2)^2}} \left(1 - e^{-\frac{\omega}{2Q}t}\right) \sin[\omega t + \phi_{\text{free}}], \tag{9.15}$$

where the phase, ϕ_{free}, is given by

$$\phi_{\text{free}} = \arctan\left[\frac{1}{Q(1/x - x)}\right], \tag{9.16}$$

and where

$$x = \omega/\omega_0. \tag{9.17}$$

We shall now present the solutions associated with the transient and steady-state regimes of the tapping and compare them with those of the vibration of the free cantilever.

9.6
Transient Regime

Figure 9.3 shows the solution to the vibration of the tapping cantilever in the transient regime, z (solid line), and to the free cantilever, z_{free} (dashed line). The difference between the two curves stems from the fact that the boundary conditions for the tapping cantilever are somewhat different than those for the free one. In contrast to operation in the noncontact mode, here the vibration of the cantilever, shown by the solid curve, is limited by tapping on the sample. Because the system is highly nonlinear, a somewhat chaotic behavior may ensue, and it will take quite a few oscillations before settling down to an almost orderly behavior. Note, however, that conditions exist under which the oscillations do become highly chaotic and do not settle down. Let us now compare z and z_{free} in the steady-state regime in the last two of the $2Q$ cycles.

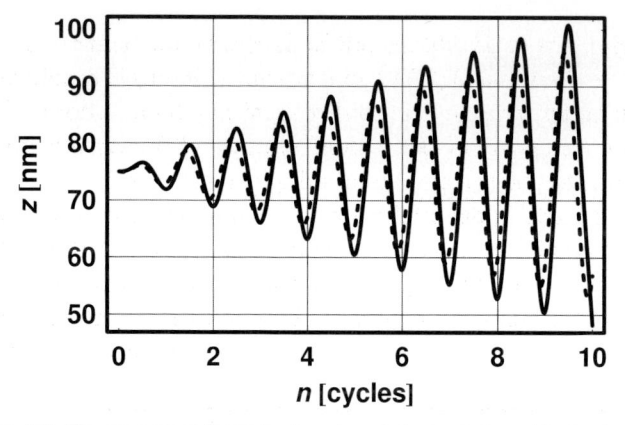

Fig. 9.3 The numerical (solid line) and analytical solutions (dashed line) for the transient vibration regime.

9.7
Steady-State Regime

A comparison of z (solid line) and z_{free} (dashed line) in the steady-state regime is shown in Figure 9.4. Here, the presence of the tip–sample force during tapping produces a shift in the phase and amplitude of the numerical solution of z relative to that of z_{free}, a topic that will be further investigated in the next sections.

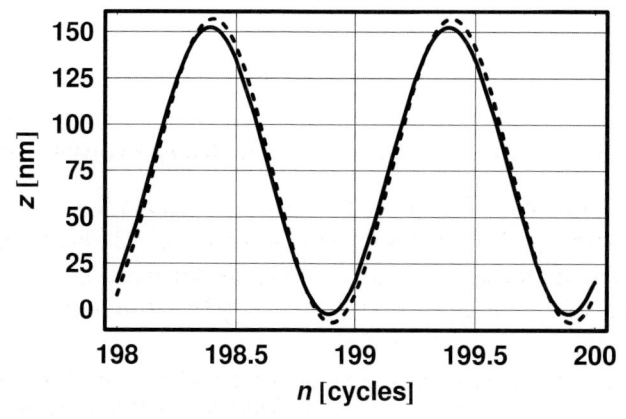

Fig. 9.4 A comparison of z (solid line) and z_{free} (dashed line) in the steady-state regime.

9.8
Tapping Phase Diagram

To further explore the phase of vibration of the cantilever, we produce a parametric plot of z against z_{free}, shown in Figure 9.5. Had z and z_{free} been in phase, the curve would have been a straight line inclined at an angle of 45°. However, the presence of a finite tip–sample force during tapping gives rise to a phase shift between these two curves as evidenced here by the elliptical shape of the plot.

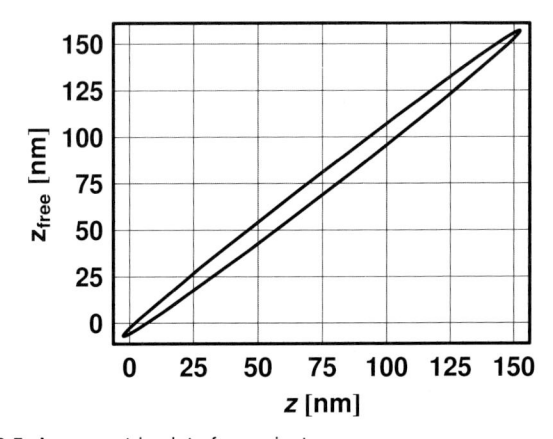

Fig. 9.5 A parametric plot of z against z_{free}.

9.9
Displacement–Velocity Phase Diagram

The displacement–velocity phase diagram is shown in Figure 9.6 for the last three cycles of the numerical calculation. Here, the parametric plot yields an almost perfect circle, indicating the good behavior of the vibration of the cantilever, namely its being an almost harmonic motion. It may happen, though, that a particular set of parameters will give rise to a chaotic behavior that will be manifested by a spiral rather than by a closed circle.

9.10
Numerical Value of the Phase Shift

A method for evaluating the phase of vibration of the tapping cantilever relative to the free one is to find the zero crossing of their velocities, v and v_{free}, respectively. Figure 9.7 shows v (solid line) and v_{free} (dashed line), respectively.

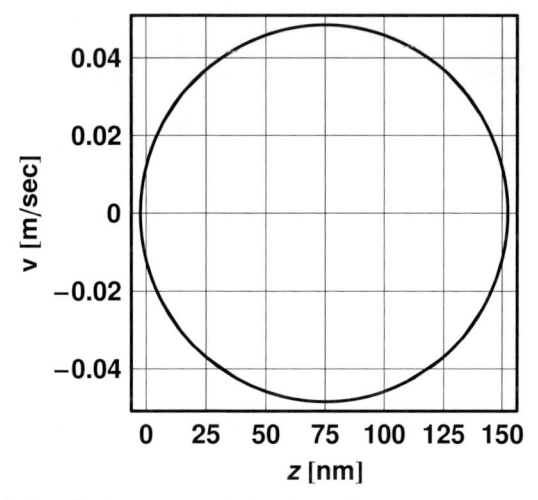

Fig. 9.6 The displacement–velocity phase diagram.

We can find the phase of vibration of the tapping cantilever, ϕ, relative to that of the free cantilever by measuring the lag between these two zero crossings.

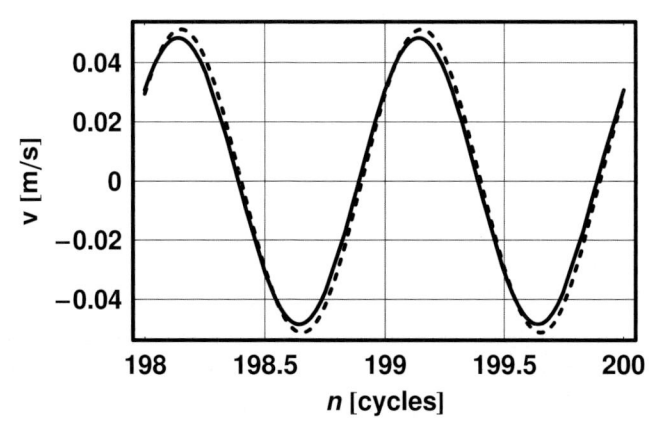

Fig. 9.7 A comparison of v (solid line) and v_{free} (dashed line) in the steady-state regime.

The displacement, z, for the last calculated cycle, shown in Figure 9.8, is found to be an almost harmonic function, with no apparent distortions due to the tapping on the sample.

The velocity, v, for the last calculated cycle is shown in Figure 9.9. The velocity, on the other hand, may show a small "glitch" when it crosses every other zero, namely this occurs when the motion of the tip is stopped by the sample.

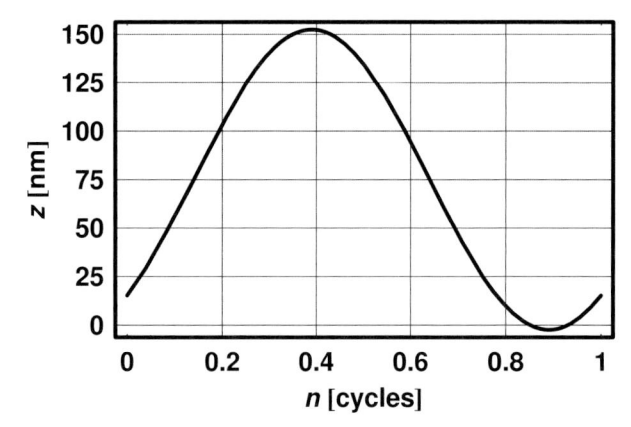

Fig. 9.8 The displacement, z, for the last calculated cycle.

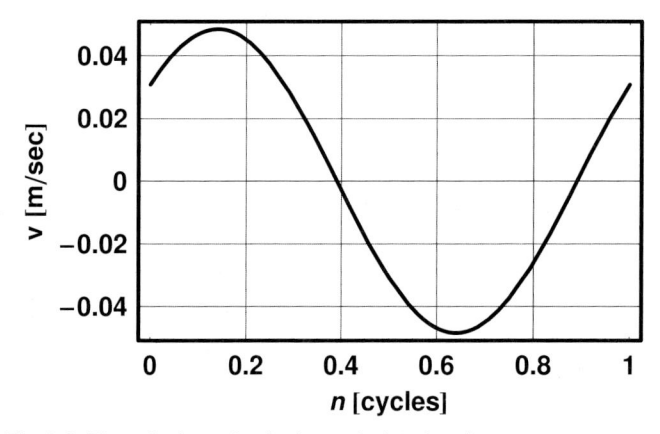

Fig. 9.9 The velocity, v, for the last calculated cycle.

Note that when operating on resonance, the tapping of the tip on the sample will decrease the amplitude of vibration, increase the average position of the tip, and shift the phase from zero to a negative value, as expected from a repulsive indentation force. However, when operating off resonance, the amplitude of vibration may increase or decrease relative to that of the free cantilever, since the tapping increases the resonance frequency of the tapping cantilever. The resultant phase may therefore be either positive or negative. The indentation, d, for the last calculated cycle is shown in Figure 9.10.

The total force, F_{total}, for the last calculated cycle is shown in Figure 9.11.

The indentation of the sample only lasts a short time relative to a full cycle, and appears as a series of pulses when observing a large number of cycles. Likewise, the force, which is proportional to the derivative of the velocity, appears as a series of pulses pointing mainly upward, since it is mostly repul-

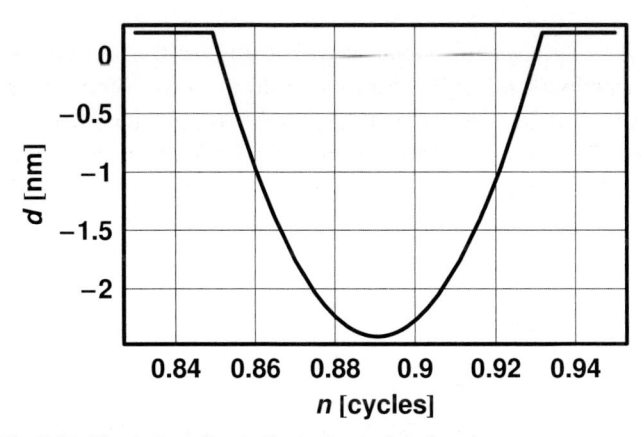

Fig. 9.10 The indentation in the last calculated cycle.

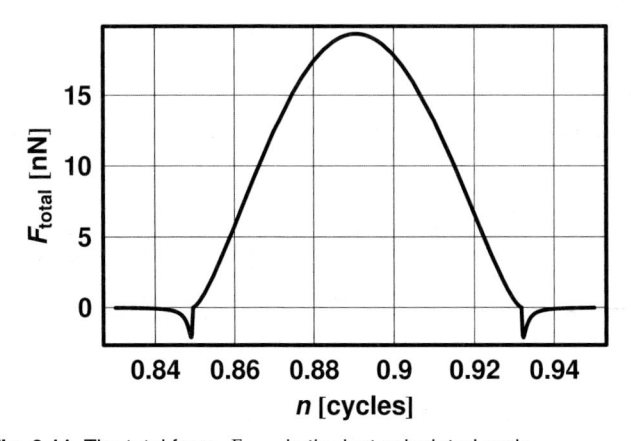

Fig. 9.11 The total force, F_{total}, in the last calculated cycle.

sive. Because of adhesion forces, there may sometimes be small spikes around both the indentation and force pulses, as observed in Figure 9.11. By changing the sample from a soft material, a polymer, to a hard one, for example silicon, one finds that the indentation of the sample by the tip becomes shallower. The indentation force, on the other hand, becomes larger, as expected. Obviously, one must take the elastic properties of the sample into account when modeling an atomic force microscope that operates in the tapping mode. It is of interest to probe the maximum radius, r_{max}, and pressure, P_{max}, of the indentation, which are given by

$$r_{max} = \left(\frac{RF_{max}}{k_s}\right)^{1/3} \quad (9.18)$$

and

$$P_{max} = \frac{3}{2\pi}\left(\frac{k_s}{R}\right)^{2/3} F_{max}^{1/3}, \qquad (9.19)$$

where F_{max} has been obtained numerically from Figure 9.11.

9.11
Summary of Results

The key results obtained in this chapter are given in Table 9.5, where the values to the left- and right-hand side of the arrows correspond to the free and tapping cantilevers, respectively.

Table 9.5 The key results obtained in this chapter.

	Free	Tapping	
f	= 99.65		kHz
A	= 82.1635	→ 77.4094	nm
z_{av}	= 75	→ 75.0072	nm
ϕ	= 0	→ −4.345 57	deg
d_{max}	=	−2.595 12	nm
r_{max}	=	5.094 23	nm
P_{max}	=	0.354 403	GPa

Note that when operating on resonance, the tapping of the tip on the sample will decrease the amplitude of vibration, increase the average position of the tip, and shift the phase from zero to a negative value, as expected from a repulsive indentation force. However, when operating off resonance, the amplitude of vibration may increase or decrease relative to that of the free cantilever, since the tapping increases the resonance frequency of the tapping cantilever. The resultant phase may, therefore, be either positive or negative.

▶ Exercises for Chapter 9

1. Calculate and plot the amplitude and phase of vibration of the cantilever for the two other frequencies, f_- and f_+, and for tapping on a Si sample.
2. Calculate and plot the case where the sample is vibrated by a bimorph, causing the cantilever to vibrate via the tip–sample force derivative.

References

1. Y. Martin, C. C. Williams, and H. K. Wickramasinghe, "Atomic force microscope force mapping and profiling on a sub 100-Å scale," J. Appl. Phys. **61**, 4723 (1987).
2. Q. Zhong, D. Inniss, K. Kjoller, and V. B. Elings, "Fractured polymer/silica fiber surface studied by tapping mode atomic force microscopy," Surf. Sci. Lett. **290**, L688 (1993).
3. H. G. Hansma and J. H. Hoh, "Biomolecular imaging with the atomic force microscope," Annu. Rev. Biophys. Biomol. Struct. **23**, 115 (1994).
4. S. N. Magonov, V. Elings, and M.-H. Whangbo, "Phase imaging and stiffness in tapping mode atomic force microscopy," Surf. Sci. Lett. **375**, L385 (1997).
5. J. P. Cleveland, B. Anczykowski, A. E. Schmid, and V. B. Ealings, "Energy dissipation in tapping-mode atomic froce microscopy," Appl. Phys. Lett. **72**, 2613, (1998).
6. J. Tamayo and R. Garcia, "Relationship between phase shift and energy dissipation in tapping-mode scanning force microscopy," Appl. Phys. Lett. **73**, 2926 (1998).
7. J. P. Hunt and D. Sarid, "Kinetics of a lossy grazing oscillator," Appl. Phys. Lett. **72**, 2969 (1998).

10
Metal–Insulator–Metal Tunneling

▎Highlights

1. General equations for metal–insulator–metal tunneling with and without an image potential
2. Tunneling current for small and large voltage approximations.
3. Barrier width and height

Abstract

The exponential dependence of electron tunneling cross-section on the thickness of the insulator of a metal–insulator–metal (MIM) structure is at the heart of scanning tunneling microscopy, because it makes tunneling sensitive to sub-angstrom changes in this thickness. The basic principles of this effect are presented in this chapter together with a set of examples that illustrate the role the different parameters play in determining the tunneling current. The model considers a MIM structure with two similar plane-parallel metal electrodes that can be readily extended to dissimilar metals. A general tunneling equation is presented, and approximate solutions for small and large voltages are given. Next, the image potential experienced by a tunneling electron in the insulator is introduced, and its effect on the tunneling current computed and compared to the small and large voltage approximations. Finally, the concept of an apparent barrier height is discussed, and the role of the image potential in lowering it is addressed.

10.1
Introduction

A tunneling current will flow between two plane-parallel metal electrodes separated by a thin insulator, when a bias is impressed between them. There is a large body of literature that addresses this important problem, in which a variety of approaches and approximations have been used. The model presented in this chapter uses the theory of Simmons [1], which yields two widely

10.1 Introduction

used equations for the tunneling current, one for the small voltage regime, relevant to scanning tunneling microscopy, and one for the large voltage regime, relevant to the so-called Fowler–Nordheim field emission. A typical energy diagram for a metal–insulator–metal structure with a triangular barrier is shown in Figure 10.1. The figure shows two similar metal electrodes having a barrier height $\phi = 4\,\text{eV}$ relative to that of the insulator. Here, the thickness of the insulator is 1 nm, and the biases across it are (a) $v = 1\,\text{V}$ and (b) $v = 5\,\text{V}$. Note that the Fermi level of the left-hand side electrode is kept at zero potential, while that of the right-hand side electrode is at a positive bias, so that the electrons tunnel to the right. In the following sections, we will find the general equations for tunneling with and without an image potential, the small and large voltage approximations for the tunneling, and obtain the effective barrier width and height. Table 10.1 shows the notations for the parameters used in this chapter. The notations for the parameters used in this chapter are given as follows:

e	positive electron charge	$\Delta s = x_1 - x_2$	effective barrier width
m_e	mass of electron in insulator	v	bias across insulator
		F	field across insulator
ϕ	metal–insulator barrier height	v_i	image potential
$\bar{\phi}$	metal–insulator mean barrier height	v	bias across insulator
		ϵ_r	relative dielectric constant of insulator
s	insulator width		
x_1 and x_2	limits of barrier at the Fermi level of metals	J	tunneling current density

The metal–insulator barrier height, ϕ, and the relative dielectric constant of the insulator, ϵ_r, are given in Table 10.1:

Table 10.1 The metal–insulator barrier height, ϕ, and the relative dielectric constant of the insulator, ϵ_r.

ϕ/e (eV)	ϵ_r
4	5

Figure 10.1 is a schematic energy diagram associated with two identical metal electrodes separated by a 1-nm-thick insulator, for an applied bias of (a) $v = 1\,\text{V}$ and (b) $v = 5\,\text{V}$. Here, x denotes the position across the structure, and M and I denote metal and insulator, respectively.

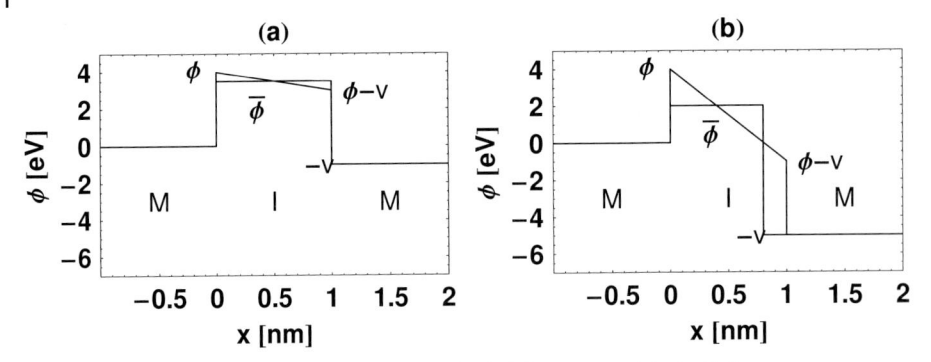

Fig. 10.1 The metal–insulator barrier height, ϕ, associated with two identical metals separated by a 1-nm-thick insulator, for an applied bias of (a) $v = 1\,\text{V}$ and (b) $v = 5\,\text{V}$. Here, x denotes the position across the structure, and M and I denote the metal and insulator, respectively.

10.2
Tunneling Current Density

10.2.1
General Solution

The general solution to the tunneling current density, J_g, across the thin insulator shown in Figure 10.1, can be obtained by using the WKB approximation,

$$J_g = \frac{e}{2\pi h \Delta s^2}\left(\bar{\phi} e^{-A\Delta s\sqrt{\bar{\phi}}} - (ev + \bar{\phi})e^{-A\Delta s\sqrt{ev+\bar{\phi}}}\right), \quad (10.1)$$

where

$$A = \frac{4\pi\sqrt{2m_e}}{h}. \quad (10.2)$$

Note that a factor β of order unity, appearing in Ref. [1], has been neglected. The general solution requires a model for the average barrier height, $\bar{\phi}$, and width, Δs, which can be approximated for two cases, one when the applied bias across the insulator is much smaller than the barrier height, and the other when it is much larger. These two approximations can then be compared to the general solution using typical parameters.

10.2.2
Small Voltage Approximation

Here we consider the case where $v < \phi/e$, as shown in Figure 10.1(a) for a bias $v = 1\,\text{V}$. One observes that the width of the barrier does not change, $\Delta s = s$, but its average height decreases to $\bar{\phi} = \phi - ev/2$. The general solution for this case, $J_{g\text{small}}$,

$$J_{g\text{small}} = J_g\left(v, s, \phi - \frac{ev}{2}\right), \tag{10.3}$$

can be approximated by J_{small},

$$J_{\text{small}} = \sqrt{2m_e\phi}\left(\frac{e}{h}\right)^2 \frac{v}{s} e^{-4\pi s \frac{\sqrt{2m_e\phi}}{h}}, \tag{10.4}$$

which is widely used in modeling scanning tunneling microscopy. The general and approximate solutions, $J_{g\text{small}}$ (solid line) and J_{small} (dashed line), shown in Figure 10.2 for $\Delta s = 1$ nm, are in good agreement for $0\,\text{V} < v < 0.8\,\text{V}$.

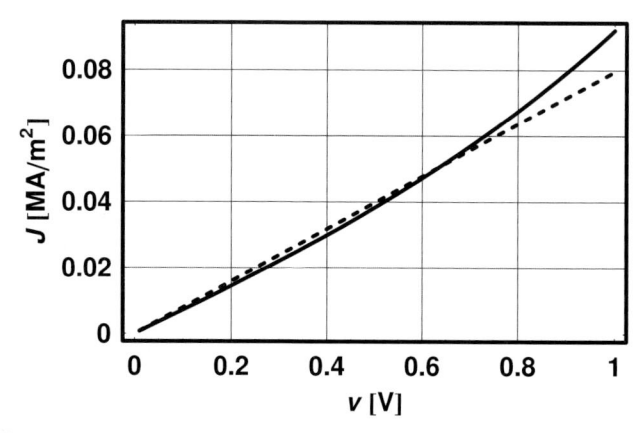

Fig. 10.2 The general and approximate solutions to the tunneling current density, $J_{g\text{small}}$ (solid line) and J_{small} (dashed line), as a function of v, for $s = 1$ nm.

A comparison of $J_{g\text{small}}$ (solid line) and J_{small} (dashed line) for $v = 1\,\text{V}$, depicted in Figure 10.3, shows an agreement in the range of $0.9\,\text{nm} < s < 1.1\,\text{nm}$.

We find that the approximate solution for the small voltage regime fits the general solution rather well, which is the reason why it has been widely used in the literature to model the operation of the scanning tunneling microscope.

10.2.3
Large Voltage Approximation

Next, consider the case where $v > \phi/e$, as shown in Figure 10.1(b), for a bias $v = 5\,\text{V}$. In this case, both the width and average height of the barrier,

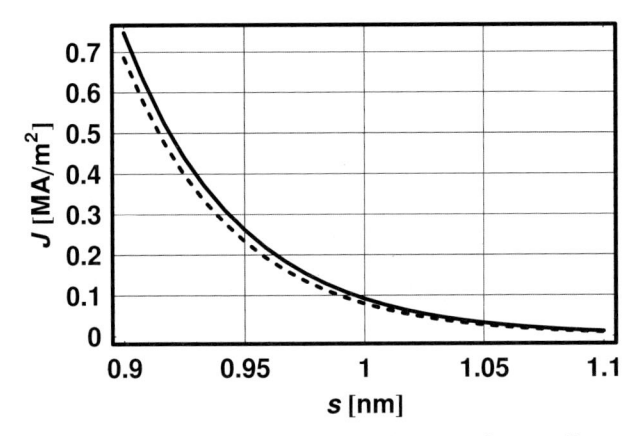

Fig. 10.3 The general and approximate solutions to the tunneling current density, $J_{g_{small}}$ (solid line) and J_{small} (dashed line), as a function of s, for $v = 1$ V.

which are functions of the applied bias, can be approximated by $\Delta s = s\phi/(ev)$ and $\bar{\phi} = \phi/2$, respectively. The general solution for this case, $J_{g_{large}}$,

$$J_{g_{large}} = J_g[v, s\phi/(ev), \phi/2], \tag{10.5}$$

can be approximated by J_{large},

$$J_{large} = \frac{e^3}{8\pi h \phi} \frac{v^2}{d^2} e^{-\frac{d}{v}\frac{8\pi\phi^{3/2}\sqrt{2m_e}}{3eh}}, \tag{10.6}$$

where $F = v/s$. Note that s does not appear explicitly in Eq. (10.6). Equation (10.5) is the so-called Fowler–Nordheim field-emission equation, which will be dealt with in the next chapter. The general and approximate solutions to the tunneling current density, $J_{g_{large}}$ (solid line) and J_{large} (dashed line), are shown in Figure 10.4, where $s = 1$ nm. Figure 10.4 shows that the two tunneling currents are in good agreement for 4 V $< v < 5$ V.

A comparison of $J_{g_{large}}$ (solid line) and J_{large} (dashed line) for $v = 1$ V, depicted in Figure 10.5, show that they are in good agreement in the range of 0.9 nm $< s < 1.1$ nm.

10.3
The Image Potential

The previous results have ignored the fact that a tunneling electron in the insulator senses an image potential, v_i, due to the presence of the two metal electrodes bounding the insulator. The image potential is given by

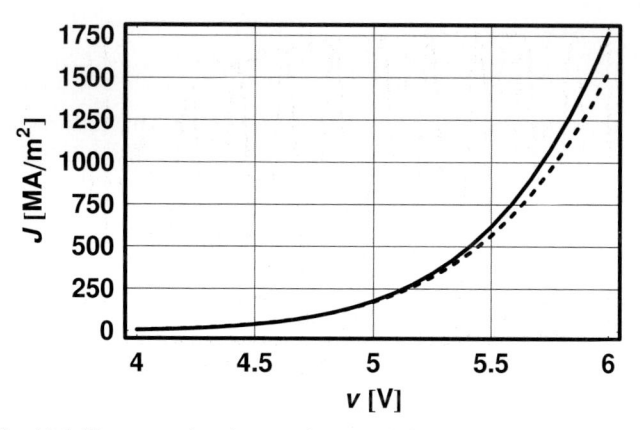

Fig. 10.4 The general and approximate solutions to the tunneling current density, $J_{g\text{large}}$ (solid line) and J_{large} (dashed line), as a function of v, for $s = 1$ nm.

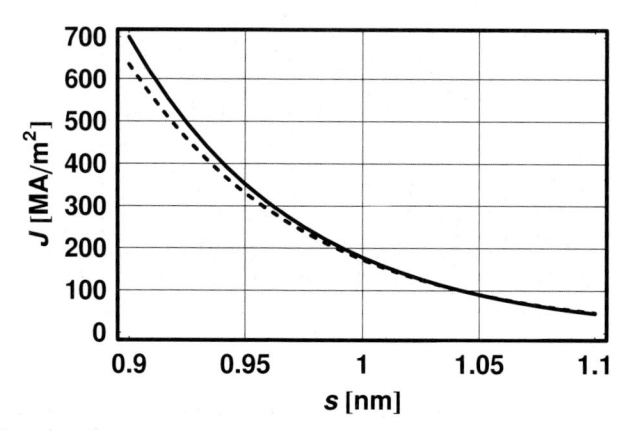

Fig. 10.5 The general and approximate solutions to the tunneling current density, $J_{g\text{large}}$ (solid line) and J_{large} (dashed line), as a function of s, for $v = 1$ V.

$$v_i = \frac{e}{4\pi\epsilon_0\epsilon_r}\left[\frac{1}{2x} + \sum_{n=1}^{\infty}\left(\frac{ns}{(ns)^2 - x^2} - \frac{1}{ns}\right)\right], \tag{10.7}$$

where x is the position of the electron inside the insulator whose width is s. A good approximation to v_i is obtained by

$$v_i = -\frac{gs}{x(s-x)}, \tag{10.8}$$

where

$$g = \frac{1.15e \ln 2}{8\pi\epsilon_0\epsilon_r}. \tag{10.9}$$

Note that the effect of the image potential is quite dramatic when the electron is in the center of the barrier, namely for $x = s/2$. The exact effect that the image potential has on the barrier height will now be presented.

10.4
Barrier with an Image Potential

10.4.1
The Barrier

The barrier height in the presence of the image potential, ϕ_i, is given by

$$\phi_i = \phi - \frac{evx}{s} + ev_i. \tag{10.10}$$

We shall first explore the dependence of ϕ_i on the bias v for $s = 1$ nm. Figure 10.6 shows ϕ_i as a function of the position inside the insulator, x, for $v = 0$ V (solid line), 3 V (dashed line), and 6 V (dotted line). One observes that increasing v, while keeping s fixed, decreases both the height and the width of the barrier.

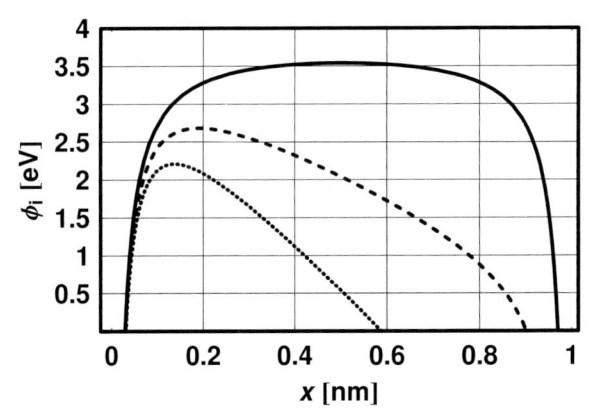

Fig. 10.6 The barrier height, ϕ_i, as a function of the position inside the insulator, x, for $s = 1$ nm, and for $v = 0$ V (solid line), 3 V (dashed line), and 6 V (dotted line).

Next, we explore the dependence of ϕ_i on the bias v for $v = 5$ V. Figure 10.7 shows ϕ_i as a function of the position inside the insulator, x, for $s = 0.9$ nm (solid line), 1 nm (dashed line), and 1.1 nm (dotted line). One observes that increasing s, while keeping v fixed, decreases both the height and the width of the barrier.

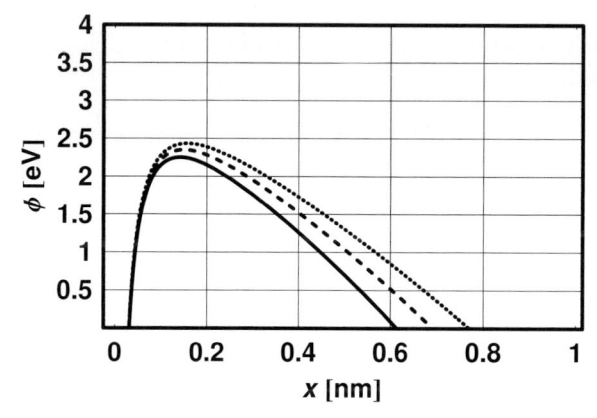

Fig. 10.7 The barrier height, ϕ_i, as a function of the position inside the insulator, x, for $v = 5\,\text{V}$, and for $s = 0.9\,\text{nm}$ (solid line), $1\,\text{nm}$ (dashed line), and $1.1\,\text{nm}$ (dotted line).

Now that we have calculated the shape of the barrier, we can go to the next step and calculate its width.

10.4.2
The Barrier Width

The width of the barrier is obtained by solving for the two zero crossings of ϕ_i as a function of v and s. The two crossings are denoted by x_{i_1} and x_{i_2}, and the barrier width by $\Delta x_i = x_{i_2} - x_{i_1}$. Figure 10.8 depicts the barrier width as a function of v, for $s = 1\,\text{nm}$.

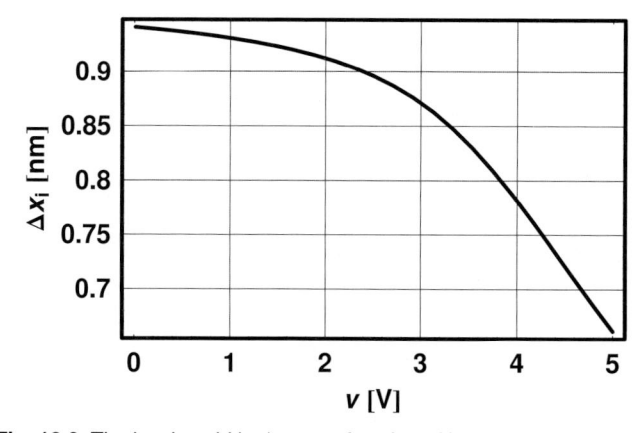

Fig. 10.8 The barrier width, Δx_i as a function of bias, v, for $s = 1\,\text{nm}$.

Figure 10.9 depicts the barrier width as a function of s, for $v = 5\,\text{V}$, showing a linear dependence on s.

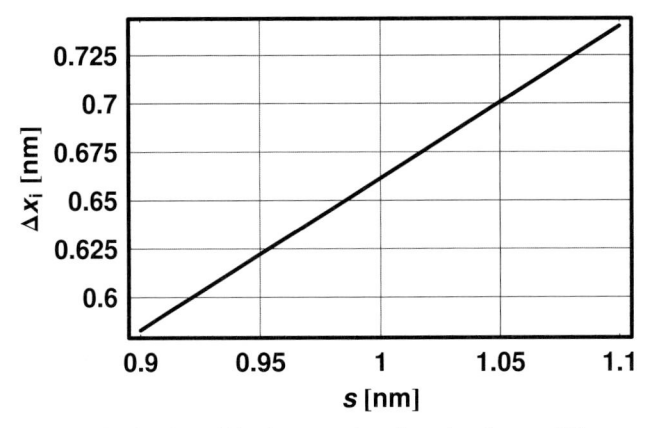

Fig. 10.9 The barrier width, Δx_i, as a function of s, for $v = 5\,\text{V}$.

These results will now be applied to the evaluation of the average barrier height, used in Eq. (10.1).

10.4.3
Average Barrier Height

The average barrier height, $\bar{\phi}_i$, is

$$\bar{\phi}_i = \frac{1}{\Delta x_i} \int_{\phi_{i_1}(v,s)}^{\phi_{i_2}(v,s)} \phi_i(v, s, x)\, dx. \tag{10.11}$$

Figure 10.10 shows the v-dependence of $\bar{\phi}_i$ for $s = 1\,\text{nm}$, where the effect of v on $\bar{\phi}_i$ is observed to be large.

Figure 10.11 shows the s-dependent average barrier height, $\bar{\phi}_i$, for $v = 5\,\text{V}$, where the effect of s is observed to be small.

10.5
Comparison of the Barriers

The barrier height, ϕ, associated with two identical metals separated by a 1-nm-thick insulator, for applied biases of (a) $v = 1\,\text{V}$ and (b) $v = 5\,\text{V}$, is shown in Figure 10.12. Here, x denotes the position across the structure, and M and I denote the metal and insulator, respectively. The figure shows the height of the barriers ϕ, ϕ_i, and $\bar{\phi}_i$, as a function of the position across the MIM structure, x, for $s = 1\,\text{nm}$ and $v = 5\,\text{V}$.

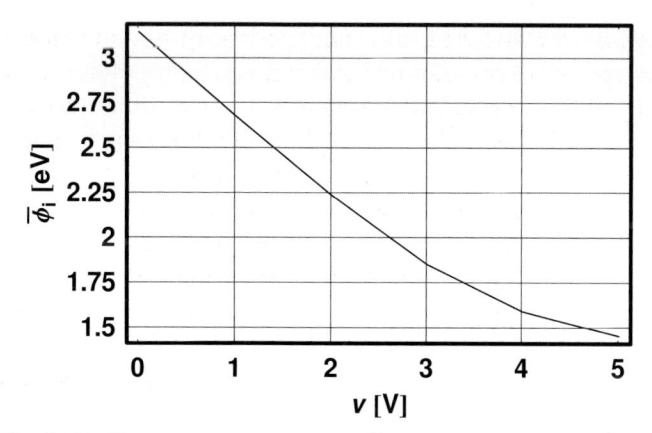

Fig. 10.10 The average barrier height, $\bar{\phi}_i$, as a function of v.

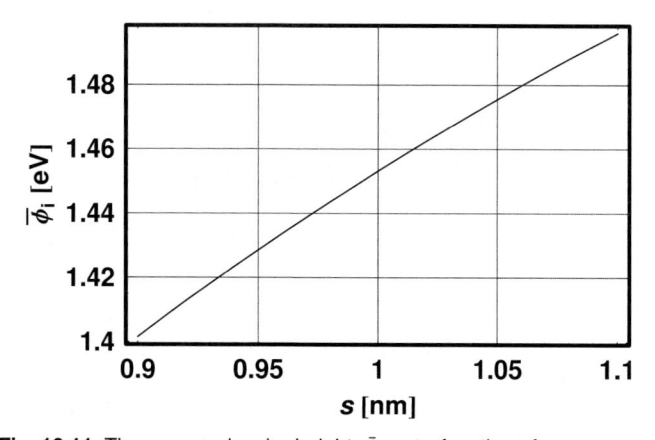

Fig. 10.11 The average barrier height, $\bar{\phi}_i$, as a function of s.

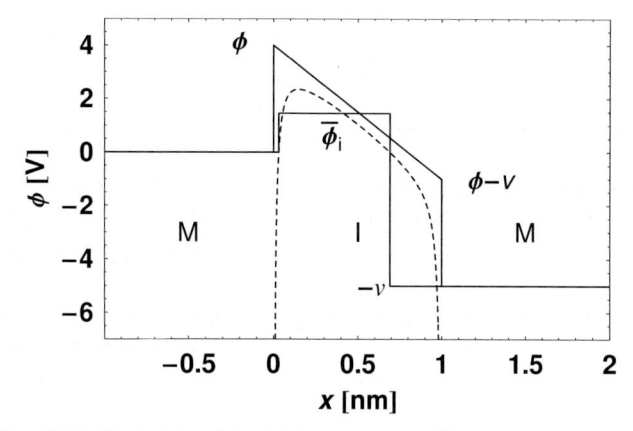

Fig. 10.12 The height of the barriers ϕ, ϕ_i, and $\bar{\phi}_i$, as a function of the position across the MIM structure, x, for $s = 1\,\text{nm}$ and $v = 5\,\text{V}$.

Now that we have found the dependence of both Δs and $\bar{\phi}_i$ on s and v, we can calculate the general tunneling current given by Eq. (10.1).

10.6
The General Solution with an Image Potential

The v- and s-dependent tunneling current density that takes into account the image potential, J_{g_i}, is obtained from the general solution, J_g, by incorporating the barrier width and average height, Δx_i and $\bar{\phi}_i$,

$$J_{g_i} = J_g(v, \Delta x_i, \bar{\phi}_i). \tag{10.12}$$

Figure 10.13 depicts the logarithm of the tunneling resistance, $\log(v/J)$, as a function of v, for v/J_{g_i} (solid line), $v/J_{g\text{small}}$ (dashed line), and $v/J_{g\text{large}}$ (dotted line), for $s = 1$ nm.

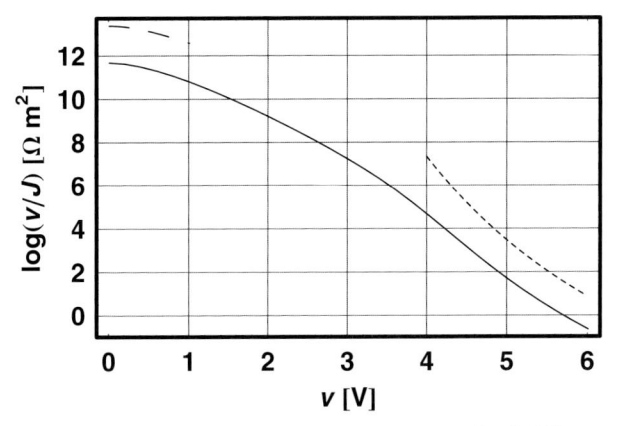

Fig. 10.13 The logarithm of the tunneling resistance, $\log(v/J)$, as a function of v, for v/J_{g_i} (solid line), $v/J_{g\text{small}}$ (dashed line), and $v/J_{g\text{large}}$ (dotted line), for $s = 1$ nm.

The dashed and dotted lines show the resistivity as a function of v, for a fixed s, in the small bias regime (left-hand side) and large bias regime (right-hand side) approximations, respectively, with no image potential. The solid line depicts the resistivity in the presence of the lowered and narrowed barrier due to the image potential. This resistivity is obviously smaller than the ones that exclude the image potential. Note that if one metal electrode is replaced by a semiconductor, then the tunneling electron in the insulator will encounter only a single rather than multiple images inside the metallic electrode. Consequently, the role of the image potential will be drastically reduced.

10.7 Apparent Barrier Height

By measuring the tunneling current as a function of s for a fixed v, in the small bias approximation, one can obtain an apparent barrier height, ϕ_{app}, as shown, for example, in Ref. [3]. To that end, we write Eq. (10.4) as

$$J_{small} = J_0 e^{-2\kappa s}, \qquad (10.13)$$

where

$$J_0 = \sqrt{2m_e \phi_{app}} \frac{e^2}{h^2} \frac{v}{s}, \qquad (10.14)$$

and

$$\kappa = \frac{2\pi \sqrt{2m_e \phi_{app}}}{h}. \qquad (10.15)$$

Here, κ is the k-vector of an electron in the insulator. Neglecting the small dependence of J_0 on s, yields

$$\kappa = -\frac{1}{2} \partial_s \ln J_{small}, \qquad (10.16)$$

from which we get

$$\phi_{app} = \frac{2}{m_e} \left(\frac{h}{4\pi} \right)^2. \qquad (10.17)$$

An important question often encountered in scanning tunneling microscopy is whether Eq. (10.17) is a good approximation if we include the effect of the image potential. To answer this question, we take the logarithmic derivative of the general tunneling equation, Eq. (10.12), where the image potential has been taken into account. Figure 10.14 is a plot of the average (solid line) and apparent (dashed line) barrier heights as a function of v, for $s = 1$ nm.

The apparent barrier height is found to be higher than the average barrier height. Figure 10.15 is a plot of the apparent barrier height as a function of s, showing that for $v = 1$ V, it decreases very little as s increases.

We found for a metal–insulator–metal structure that the apparent barrier height, ϕ_{app}, is higher than the average barrier height, $\bar{\phi}_i$. For a metal–insulator–semiconductor structure, however, the apparent barrier height is much closer to the average barrier height because the image potential in this case can be practically neglected. Note also that the results obtained here assumed plane-parallel conducting electrodes. For scanning tunneling microscopy, where one of the electrodes is a sharp tip, the electric fields are not uniform, resulting in a modified image potential and therefore a modified

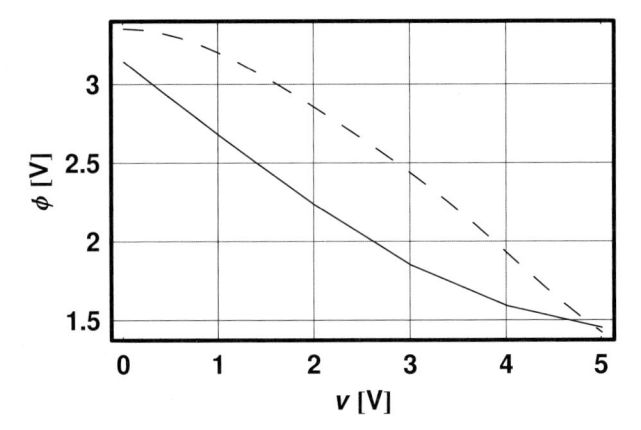

Fig. 10.14 The average (solid line) and apparent (dashed line) barrier height, ϕ, as a function of v.

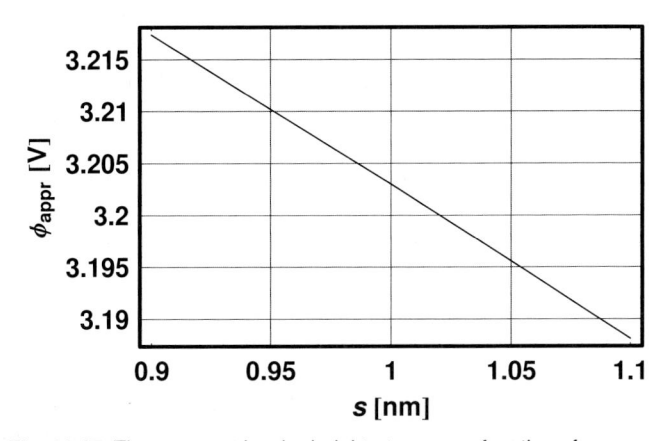

Fig. 10.15 The apparent barrier height, ϕ_{app}, as a function of s.

shape of the barrier. The results obtained in this section can serve as a guide when using the calculated value of $\partial \ln i / \partial s$ to determine the apparent barrier height.

▶ **Exercises for Chapter 10**

1. Extend the model to two dissimilar metals.
2. Consider a spherical STM tip and a plane metal substrate.
3. Use the tip–sample electric field distribution to calculate the barrier height.
4. Show that it is easier to tunnel out of the tip of an STM than into it.

References

1. J. G. Simmons, "Generalized formula for the electric tunnel effect between similar electrodes separated by a thin insulating film," J. Appl. Phys. **34**, 1793 (1963).
2. J. G. Simmons, "Electric tunnel effect between dissimilar electrodes separated by a thin insulating film," J. Appl. Phys. **34**, 2581 (1963).
3. J. H. Coombs, M. E. Welland, and J. B. Pethica, "Experimental barrier height and the image potential in scanning tunneling microscopy," Surf. Sci. Lett. **198**, L353 (1988).

11
Fowler–Nordheim Tunneling

■ **Highlights**

1. Tunneling current density
2. Oxide field and applied field
3. Tunneling oscillations and their average

Abstract

Fowler–Nordheim (FN) field emission, or tunneling, of electrons through a metal–oxide–semiconductor (MOS) structure is a powerful tool used to characterize oxide films whose thickness is on the order of several nanometers. A schematic band diagram of a biased MOS structure is presented, together with a list of notations for the various parameters used in the modeling. Next, the tunneling equation is discussed in detail and plotted by using typical experimental parameters. Oscillations in the tunneling current due to resonance effects of electrons traveling in the conduction band of the oxide are computed and plotted, first using a simple monoenergetic beam of electrons, and second by averaging over all the relevant electronic energies.

11.1
Introduction

Figure 11.1 is a schematic band diagram of a metal–oxide–silicon system in the Fowler–Nordheim (FN) bias range. The grounded metal and biased silicon electrodes, shown on the left and right-hand sides of the figure, respectively, are separated by a nm-thick silicon oxide. An externally applied bias, v_{app}, tilts the valence and conduction bands of the oxide downward so that the distance the electrons travel in their conduction band, L_{cb}, becomes finite, defining the onset of the FN tunneling regime. The electrons have real wave vectors in the metal and in the silicon. In the oxide, however, the wave vector of the electrons is real only in the region denoted by L_{cb} and imaginary in the rest of the oxide. Note that the Fermi level of the silicon in Figure 11.1 is shown as being in between the valence and conduction bands. The definitions of the parameters used in this chapter are given as follows:

e	positive electron charge	k	real wavevector of electrons in the silicon
ϕ_{Si}	workfunction of silicon		
ϕ_m	workfunction of metal	E_{F_m}	metal Fermi level
m_{eff}	effective electron mass in the oxide	E_{F_s}	silicon Fermi level
		E_{cb}	energy of the conduction band of silicon
v_{app}	externally applied bias		
v	bias across the oxide	E_{vb}	energy of the valence band of silicon
d	oxide layer thickness	E_n	energy of electrons below the metal Fermi level
F	field across the oxide		
κ	imaginary wave vector of electrons in the oxide	L_{cb}	the distance electrons travel in the oxide conduction band
κ_1	real wavevector of electrons in the oxide	$\delta\phi$	metal-silicon barrier height difference

Table 11.1 gives typical values of barrier heights of selected materials, ϕ, useful for the modeling presented in this chapter.

Table 11.1 Typical values of the barrier height, ϕ, of selected materials.

Material	ϕ (eV)
Au	4.2
Si	4.2
Cr	3.51
Al	3.17
Pt	3.01
W	3.5
Mg	2.3

11.2 Fowler–Nordheim Current Density

In the previous chapter we treated the case of tunneling current densities flowing between two plane-parallel metal electrodes separated by a thin insulator,

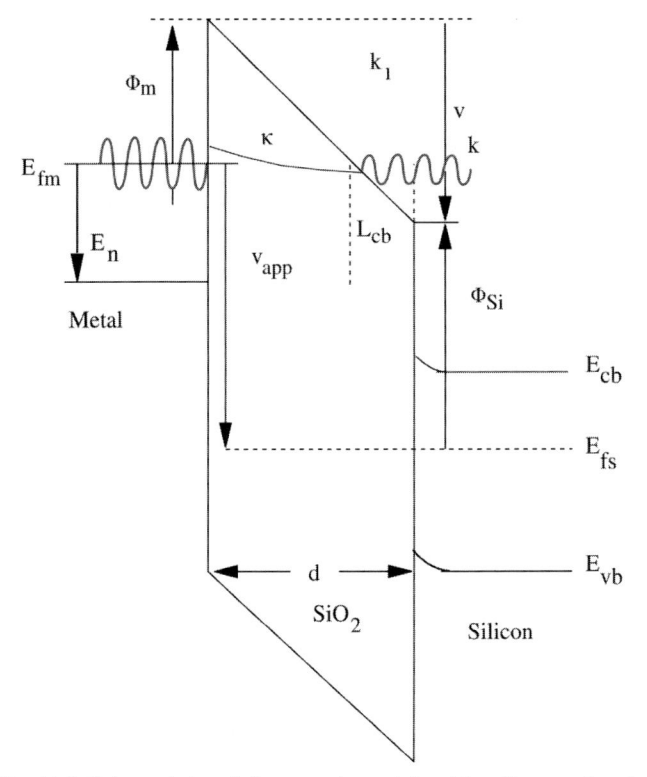

Fig. 11.1 Schematic band diagram of a metal–oxide–silicon system in the FN bias range.

in the large voltage regime. We will now address the tunneling currents in a MOS structure under similar bias conditions. Here, the tunneling current density, J,

$$J = B_0 F^2 e^{-C_0/F}, \qquad (11.1)$$

is given in terms of the field across the oxide, $F = v/d$. The two parameters B_0 and C_0 in Eq. (11.1) can be obtained from experimentally derived properties of the three media composing the MOS structure. The FN tunneling current density, as given by Eq. (11.1), will now be discussed in some detail and typical values of the parameters B_0 and C_0 presented. We start by considering the imaginary wave vector of an electron in the oxide layer, κ, which, in the parabolic approximation, is given by

$$\kappa = \frac{2\pi \sqrt{2 m_{\text{eff}} \phi}}{h}. \qquad (11.2)$$

The WKB approximation yields the coefficient C_0 of Eq. (11.1) as an averaged wave vector,

$$C_0 = \frac{2}{e}\int_0^\phi \kappa\, d\phi. \tag{11.3}$$

The integration gives an explicit expression for C_0 in terms of the effective mass of the tunneling electrons, m_{eff}, and the barrier height, ϕ,

$$C_0 = \frac{8\pi\phi^{3/2}\sqrt{2m_{\text{eff}}}}{3eh}. \tag{11.4}$$

The same approximation also gives the coefficient B_0 of Eq. (11.1),

$$B_0 = \frac{e^3}{8\pi\hbar\phi}. \tag{11.5}$$

Equation (11.4) and Eq. (11.5) can now be inserted into Eq. (11.1), yielding the tunneling current density J,

$$J = \frac{e^3}{8\pi\phi h}\frac{v^2}{d^2}e^{-\frac{d}{v}\frac{8\pi\phi^{3/2}}{3eh}\sqrt{2m_{\text{eff}}}}, \tag{11.6}$$

which is identical to Eq. (11.6) in the previous chapter. To evaluate Eq. (11.6), one needs to use the parameters associated with the particular choice of a MOS structure which will be given in the next section.

11.3
Numerical Example

As an example, consider an oxide layer with a thickness $d = 3\,\text{nm}$ and area $A = 100\,\text{nm}^2$, which yield a tunneling current of 10^5 pA when a bias of 8 V is applied across it. The applied bias differs from the bias across the oxide by about 1 V, because of the different Fermi levels of the metal and the silicon. According to Eq. (11.6), this choice of current leaves two unknowns, ϕ and m_{eff}. Since Eq. (11.5) is exponential in both, we have to be quite careful in the choice of one to determine the other. Suppose we know that $m_{\text{eff}} = 0.8\,m_e$, where m_e is the electronic mass. The metal–oxide barrier height, ϕ_x, which yields $J = 10^5\,\text{A/m}^2$, given in Table 11.2, is reasonable for the chosen parameters.

A plot of J for a bias in the range of $5\,\text{V} < v < 8\,\text{V}$ is depicted in Figure 11.2, for the parameters given in Table 11.2. Note the limited range of bias voltages that is experimentally accessible before the current becomes too large, damaging the oxide film.

Table 11.2 The experimental metal–oxide barrier height, ϕ_x, such that $J = 10^5$ A/m^2.

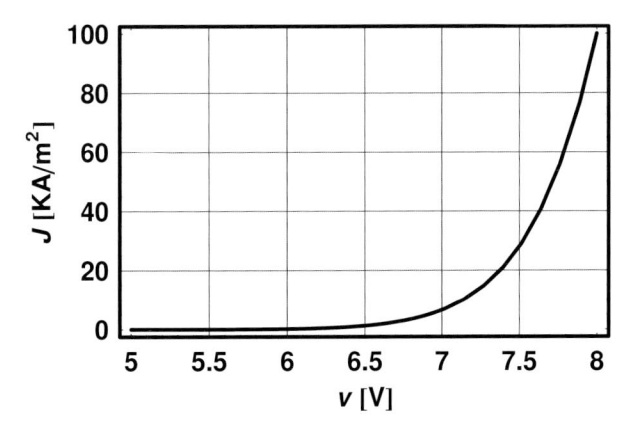

$m_{\text{eff}} = 0.8\, m_e$
$d = 3$ nm
$A = 100$ nm^2
$v = 8$ V
$J_x = 10^5$ A/m^2
$\phi_x = 3.829$ eV

Fig. 11.2 The current density, J, as a function of the bias, v.

11.4 Oxide Field and Applied Field

Let the metal be grounded so that its potential is zero, and let the external bias be applied to the silicon side of the MOS structure. There is a built-in potential between the metal and the silicon, $\delta\phi$, due to the difference in their barrier heights,

$$\delta\phi = \phi - \phi_{\text{Si}}, \tag{11.7}$$

which the externally applied bias, v_{app}, must first overcome. The actual bias across the oxide, v, is therefore given by

$$v = v_{\text{app}} - \delta\phi. \tag{11.8}$$

The field across the oxide, F, in terms of the applied bias, v_{app}, is now given by

$$F = v/d. \tag{11.9}$$

11.5
Oscillation Factor

The theoretical and experimental observations of oscillations in the tunneling current as a function of applied voltage have been discussed in the literature, and modeled by assuming a purely trapezoidal barrier without image charge effects. Despite the crude assumptions involved in this theory, the experimental data were shown to fit it quite well. The oscillating FN tunneling current density, J_{osc}, is given as a product of J and a multiplicative factor b,

$$J_{osc} = bJ(v, E_n), \tag{11.10}$$

where E_n is the energy of the tunneling electrons relative to the Fermi level of the metal. The factor b, for monoenergic electrons and for a constant E_n, is given by

$$b = \frac{b_0}{\text{AiryAi}(-k_1 L_{cb})^2 + \left(\frac{k_1}{k}\right)^2 \text{AiryAi}'(-k_1 L_{cb})^2}. \tag{11.11}$$

Here, AiryAi and AiryAi' are the Airy function and its derivative, respectively, and $b_0 = 0.08$ is an arbitrary factor that normalizes b so that its average is around unity. The wave vector of electrons propagating in the conduction band of the oxide, k_1, is given by

$$k_1 = \left(\frac{8\pi^2 m_{\text{eff}} ev}{dh^2}\right)^{1/3}. \tag{11.12}$$

The distance the electrons propagate in the conduction band of the oxide, L_{cb}, calculated geometrically for the trapezoidal barrier of Figure 11.1, is

$$L_{cb} = d - \frac{(\phi + E_n)}{e} \frac{d}{v}. \tag{11.13}$$

Here, E_n is the energy of the tunneling electrons relative to the metal's Fermi energy E_F, such that E_n is positive for electrons with energy below E_F. The wave vector of electrons propagating in the silicon, k, is given by

$$k = \frac{2\pi \sqrt{2m_{\text{eff}}(\phi + ev - E_{cb})}}{h}. \tag{11.14}$$

Table 11.3 gives the values of k_1, k, and L_{cb} for a bias of $v = 8\,\text{V}$ applied across the oxide.

Figure 11.3 shows $b(v, 0)$ (solid line) and $b(v, 0.2\,\text{eV})$ (dashed line) as a function of v. Note that for a larger value of E_n, the electrons have less energy and will therefore require a higher bias to express a given resonance. The oscillating current, J_{osc}, can now be written as a product of two factors.

Figure 11.4 depicts J (solid line) and J_{osc} (dashed line) as a function of v, demonstrating that their difference is large enough to be observed experimentally.

Table 11.3 The values of k_1, k, and L_{cb}, for a bias of $v = 8\,\text{V}$ applied across the oxide.

$$k_1 = 3.825\,71\,1/\text{nm}$$
$$k = 13.461\,1/\text{nm}$$
$$L_{cb} = 1.563\,95\,\text{nm}$$

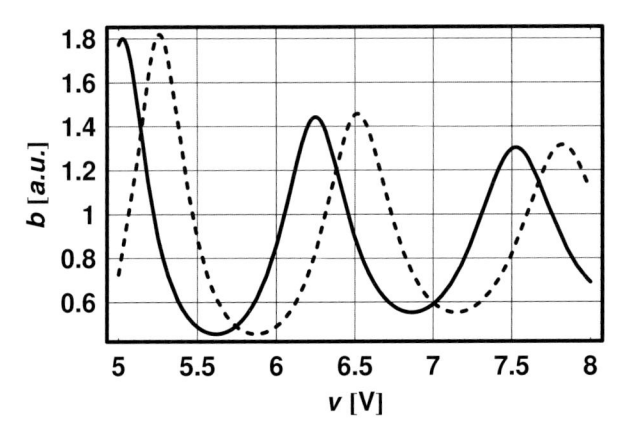

Fig. 11.3 The oscillations of $b(v,0)$ (solid line) and $b(v, 0.2\,\text{eV})$ (dashed line) as a function of v.

Instead of using the function b to describe the current oscillations, one may use the function G given by

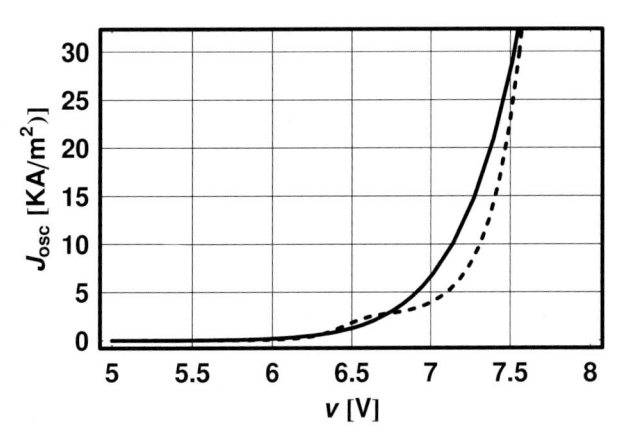

Fig. 11.4 The current densities J (solid line) and J_{osc} (dashed line) as a function of v.

$$G = -\frac{v}{d} \ln\left(\frac{J_{osc}(v, 0)}{B_0 \left(\frac{v}{d}\right)^2}\right), \quad (11.15)$$

which is shown in Figure 11.5 in terms of $F = v/d$. By choosing a different set of parameters, one observes that the oscillations that go through local minima and maxima are very sensitive not only to the oxide thickness but also to the effective mass of the tunneling electrons and to the height of the metal–oxide barrier.

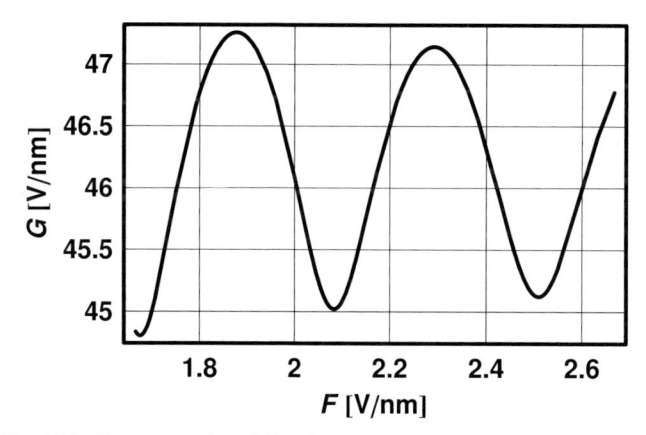

Fig. 11.5 G as a function of F, using a parametric plot.

11.6
Averaged Oscillations

To calculate actual tunneling currents, we must integrate b over all the energies for which oscillations take place. This weighted average $\bar{b}$ (see Figure 11.6) is given by

$$\bar{b} = C_2 \int_0^{ev - \phi_{Si}} b(v, E_n) e^{-C_2(\phi, v) E_n} \, dE_n, \quad (11.16)$$

where C_2 is given by

$$C_2 = \frac{C_0 d}{\phi v}. \quad (11.17)$$

Note that the integration is from the metal Fermi level to the bottom level of the oxide band edge. The averaging is responsible for the decay in the oscillation amplitude at larger fields. Note that for a given value of E_n, $\bar{b}$ will differ somewhat from b in both amplitude and periodicity. Since the exact parameters involved in the modeling of these two oscillating functions are not known exactly, one may also use $\bar{b}$ for a comparative study of different MOS structures.

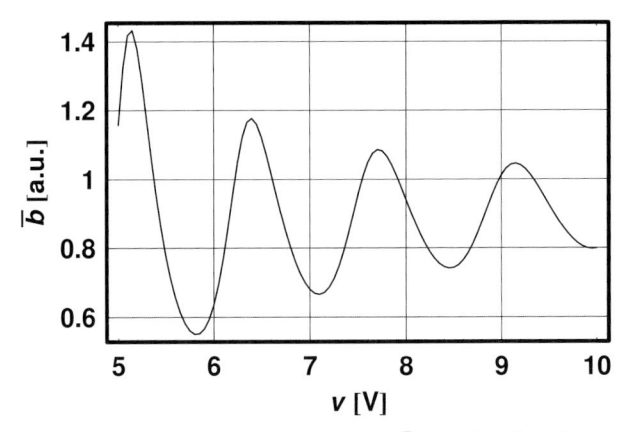

Fig. 11.6 The averaged current oscillations, $\bar{b}$, as a function of v.

11.7
Effective Tunneling Area

We have calculated in this chapter the density of the tunneling current. Experimentally, however, one measures the current itself, so the effective tunneling area has to be determined. For a plane-parallel structure, where the diameter of the electrodes is much larger than the oxide thickness, it is straightforward to figure out this area. If, however, the metal electrode is the conducting tip of an atomic force microscope, then one has to resort to a more complicated procedure. The simplest method for estimating the effective contact area is obviously to associate it with the smallest resolvable feature in a FN map, obtained by scanning the tip across a sample, and measuring the tip–sample bias required to maintain a constant tunneling current. It is also possible to model the effective area using the contact area between a deformable metallic sphere representing the tip and the deformable oxide surface. To that end, one has to measure the force acting between the tip and the sample with a conducting tip AFM. To calculate the effective area, A_{eff}, we consider only FN emission from that part of the tip that is in contact with the oxide surface. The tip–sample contact radius, r_c, is given by

$$r_c^3 = \frac{3}{4}(\kappa_1 + \kappa_2) F_{\text{ts}} R_{\text{tip}}, \qquad (11.18)$$

as shown in a previous chapter. Here, F_{ts} is the tip–sample force, R_{tip} the radius of the tip apex, and

$$\kappa_i = \frac{1 - \nu_i}{E_i}, \qquad (11.19)$$

where E_i and ν_i are Young's moduli and Poisson's ratios of the tip and sample, respectively. The effective area for tunneling is then given by

$$A_{\text{eff}} = \pi r_c^2. \tag{11.20}$$

For example, a tungsten tip with $R_{\text{tip}} = 50\,\text{nm}$ and a force of several hundreds of nN yields an approximate contact area with a silicon oxide surface of the order of $10\,\text{nm}^2$. It is clear that a meaningful FN map can only be obtained if the effective area remains constant along the scanning area, a situation that can be approximated by operating the AFM in both constant force and constant current modes.

▶ Exercises for Chapter 11

1. Use the model with other relevant parameters and explore the meaning of the computational results.
2. Test experimental FN current densities reported in the literature using the above theory.
3. Calculate effective contact areas of a sphere–plane system using relevant elastic properties and forces encountered in atomic force microscopy.

References

1. J. Maserjian, "Tunneling in thin MOS structures," J. Vac. Sci. Technol. **11**, 996 (1974).
2. G. P. Peterson, C. M. Stevenson, and J. Meserjian, "Resonance effects at the onset of FN tunneling in thin MOS structures," Solid State Electron. **18**, 449, (1975).
3. M. P. Murell, M. E. Welland, S. J. O'Shea, T. M. Wong, J. R. Barnes, and A. W. McKinnon, "Spatially resolved electric measurements of SiO_2 gate oxides using atomic force microscopy," Appl. Phys. Lett. **62**, 786 (1993).
4. T. G. Ruskell, R. K. Workman, D. Chen, D. Sarid, S. Dahl, and S. Gilbert, "High-resolution FN field emission maps of thin silicon oxide layers," Appl. Phys. Lett. **68**, 93 (1996).

12
Scanning Tunneling Spectroscopy

Highlights

1. Plots of i, $\partial i/\partial v$, and $\partial \ln i/\partial \ln v$ as a function of v
2. Spectroscopy of C_{60} molecules

Abstract

Scanning tunneling microscopy (STM) has been a powerful tool in mapping local density of states (LDOS) across conducting surfaces. It has been shown that the logarithmic derivative of the tunneling current with respect to the tip–sample voltage, termed scanning tunneling spectroscopy (STS), yields an LDOS that is practically independent of the tip–sample distance. The logarithmic derivative, which is a unitless quantity, has been widely used to generate STS data. The code in this chapter can be used to process raw STM tunneling current vs. voltage data, presented in the form of a list, to generate plots of i, $\partial i/\partial v$, and $\partial \ln i/\partial \ln v$ as a function of v. As an example, the code uses experimental $i(v)$ data, obtained by an ultrahigh vacuum STM, of C_{60} molecules chemisorbed on a Si(100)–2 × 1 surface, and compares it to the theory presented in this chapter.

12.1 Introduction

Tunneling phenomena in metal–insulator–semiconductor (MIS) structures differ from those obtained in metal–insulator–metal (MIM) structures, because of (a) the absence of an image potential, (b) the presence of surface states, and (c) the modification of the semiconductor electronic structure in the presence of a metal tip. As a result of these differences, $i(v)$ data obtained by STS contain a wealth of information about the surface of a probed semiconductor that requires special processing. The following sections deal with the analysis of tunneling current data that lead to an STS curve reflecting the LDOS of the probed semiconductor. This chapter provides a code that processes a list of experimentally obtained current vs. voltage data points, generating plots of i, $\partial i/\partial v$, and $\partial \ln i/\partial \ln v$. The logarithmic derivative is shown to reflect the LDOS of the probed surface of the semiconductor, as described in Ref. [1] to Ref. [3], while Ref. [4] is the source of the experimental results used in the examples. A schematic diagram of a scanning tunneling microscope, shown in Figure 12.1, depicts an atomically sharp tip mounted on a piezoelectric tube (PZT), separated by a subnanometer spacing from C_{60} molecules deposited on a sample. The application of a tip–C_{60} molecule voltage gives rise to a tunneling current between these two.

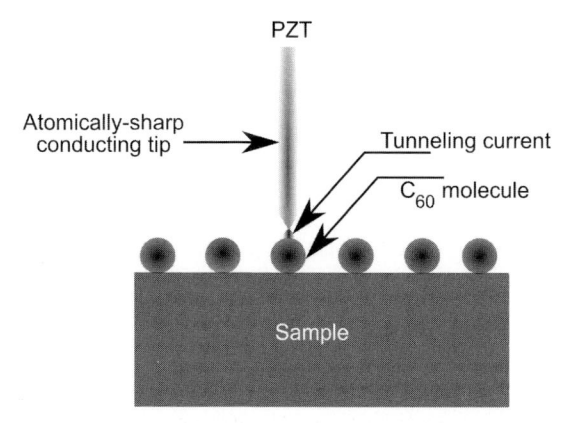

Fig. 12.1 A schematic diagram of a scanning tunneling microscope.

A general equation for the tunneling current, i, as a function of the tip–sample voltage, v, is given by

$$i = \frac{4\pi e}{\hbar} \int_{-\infty}^{\infty} d\mathcal{E} |M_{m,n}|^2 [f_F(E_F - ev + \mathcal{E}) - f_F(E_F + \mathcal{E})] \rho(\mathcal{E}, E_F), \quad (12.1)$$

where E_F is the Fermi level, $\mathcal{E}$ denotes the energy of the tunneling electrons, and M is the tip–sample tunneling matrix element, given by

$$M_{m,n} = \frac{\hbar}{2m_e} \int_{\Sigma} (\chi_m^* \nabla \psi_n - \psi_m^* \nabla \chi_n) \, dS. \quad (12.2)$$

Here, ψ and χ are the tip and sample wavefunctions, respectively, and the integration is on a surface, Σ, separating the tip and sample. Also, the Fermi function, f_F, given by

$$f_F = \frac{1}{1 + e^{\frac{\mathcal{E} - E_F + ev}{K_B T}}}. \quad (12.3)$$

where K_B is Boltzmann's constant, is depicted in Figure 12.2.

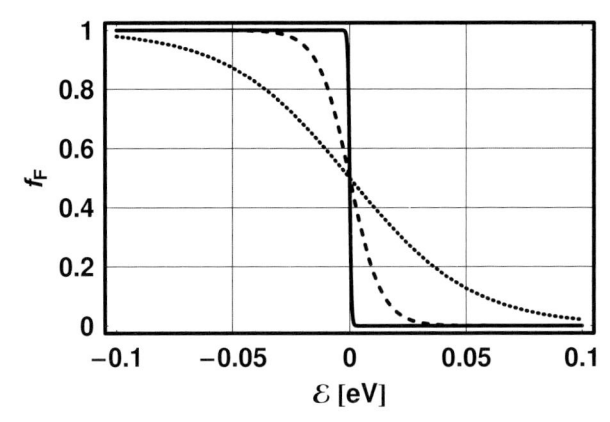

Fig. 12.2 The Fermi function, f_F, for $T = 4.2\,\text{K}$ (solid line), 77 K (dashed line), and 300 K (dotted line).

The total density of states of the system, composed of the STM tip and the surface of the sample, ρ, is the product of their respective densities of states, ρ_t and ρ_s, and is given by

$$\rho(\mathcal{E}, E_F, v) = \rho_s(\mathcal{E}, E_F, v)\rho_t(\mathcal{E}, E_F, v). \tag{12.4}$$

12.2
Fermi–Dirac Statistics

We shall first assess the role the Fermi–Dirac statistics plays in limiting the resolution of STS measurements at different temperatures. We will use the three experimentally convenient temperatures – liquid helium (4.2 K), liquid nitrogen (77 K), and ambient (300 K) – to demonstrate the behavior of the Fermi function, Eq. (12.3). Figure 12.2 shows the Fermi function, f_F, for each of these three temperatures. Here, the sharpest distribution is obtained at lowest temperature. The tip–sample bias voltage, v_b, determines the window opened for tunneling between $\mathcal{E} = 0$ and $\mathcal{E} = v_b$, while the Fermi function determines its sharpness. Only states within this window contribute to the tunneling current, albeit with unequal weight.

Figure 12.3 shows the window for $v_b = 1\,\text{V}$ under ambient conditions. Note that at $T = 0$, the window will have absolutely sharp side walls. Note also that at low enough temperatures, the width of the side walls can be approximated by $\Delta\mathcal{E} = 4K_B T$. At elevated temperatures, however, the width of the side walls will be considerably larger. This width can be obtained directly by convoluting the Fermi function with itself, yielding the function f_c,

$$f_c(\epsilon, E_F, T) = f_F(\epsilon, E_F, T)[1 - f_F(\epsilon, E_F, T)]. \tag{12.5}$$

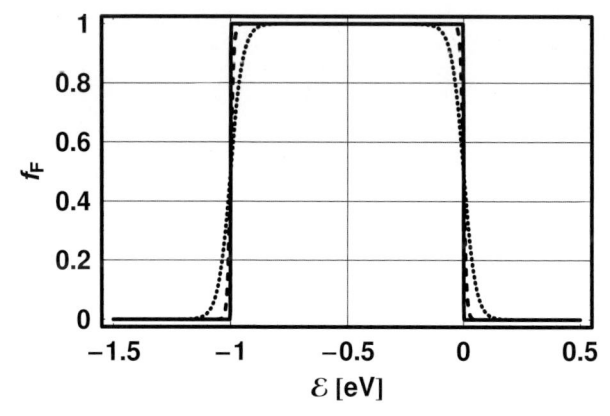

Fig. 12.3 The window determined by the Fermi function, f_F, for $T = 4.2\,K$ (solid line), $77\,K$ (dashed line), and $300\,K$ (dotted line).

The convolution function, depicted in Figure 12.4, for the three temperatures used before, demonstrates that the narrowest distribution, shown by the solid line, is obtained at the lowest temperature. For ambient conditions, however, the width of the side walls is roughly 0.1 eV, limiting the energy resolution of the STM.

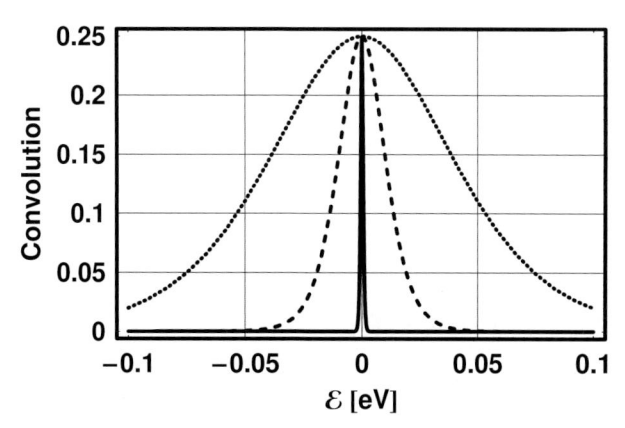

Fig. 12.4 The convolution function for $T = 4.2\,K$ (solid line), $77\,K$ (dashed line), and $300\,K$ (dotted line).

12.3 Feenstra's Parameter

Assume now that experimentally obtained STS features have a width larger than that which is determined by the Fermi function. Under such conditions, one can neglect the Fermi functions and approximate the tunneling current, i, by

$$i \propto \frac{4\pi e}{\hbar} \int_0^{ev} |M|^2 \rho(\mathcal{E}, E_F, v) \, d\mathcal{E} \,. \tag{12.6}$$

A further simplification occurs if the matrix elements and the tip density of states are independent of v. In this case the tunneling current takes an even simpler form,

$$i \propto \frac{4\pi e}{\hbar} \int_0^{ev} \rho(\mathcal{E}, E_F, v) \, d\mathcal{E} \,. \tag{12.7}$$

The sample density of states, ρ_s, can now be readily obtained from the tunneling current by taking the derivative of the current with respect to the tip–sample voltage,

$$\rho_s \propto \frac{\partial i}{\partial v} \,. \tag{12.8}$$

In practice, however, the matrix element depends on both the tip–sample bias and distance, s. As a result, plotting Eq. (12.8) as a function of voltage yields curves that depend on s. A good approximation to the density of states of the sample, which yields s-independent STS features, was proposed by Feenstra. This approximation is given by the logarithmic derivative of the tunneling current with respect to v,

$$\rho_s \propto \frac{\partial \ln i}{\partial \ln v} \,. \tag{12.9}$$

It was found experimentally that Eq. (12.9) works well for metals and narrow-gap semiconductors, as discussed in Ref. [1].

12.4
Scanning Tunneling Spectroscopy

12.4.1
STS Data File

To demonstrate the utility of Eq. (12.9), we present experimental $i(v)$ data, obtained by an ultrahigh vacuum STM, of C_{60} molecules chemisorbed on a Si(100)–2 × 1 surface, and compare it to the theory presented in this chapter. The file contained in the code accompanying this chapter is in the form of a list of N current points, i, for a voltage range of $-3\,\text{V} < v < 3\,\text{V}$. Table 12.1 gives the value of N, the minimum and maximum values of the current (in arbitrary units), $i_{\min}$ and $i_{\max}$, respectively, and the maximum difference $\Delta i_{\max}$.

12.5 Spectroscopic Data

Table 12.1 Statistics of the experimentally obtained $i(v)$ data points.

$N = 240$
$i_{min} = -102\,434\,386$
$i_{max} = 107\,218\,587$
$\Delta i_{max} = 209\,652\,973$

12.4.2
Data Processing

Because experimental data are usually noisy, we chose to perform a moving average across five data points. It is also convenient to normalize the averaged data and get a new set of current data whose range is around unity, as given in Table 12.2, with the same notation as that of Table 12.1.

Table 12.2 Statistics of the averaged $i(v)$ data list.

$N = 240$
$i_{min} = -0.611\,91$
$i_{max} = 0.388\,09$
$\Delta i_{max} = 1$

12.5
Spectroscopic Data

Figure 12.5 shows the current, i, as a function of v, after performing a moving average across five data points.

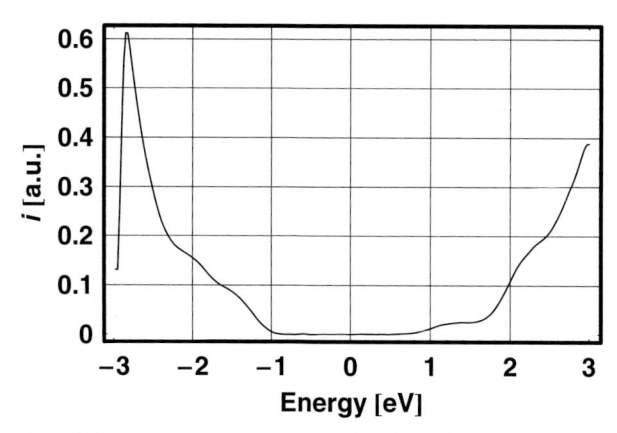

Fig. 12.5 The averaged current, i, as a function of v.

The current derivative, $\partial i/\partial v$, as a function of v, obtained after performing a moving average across 11 data points, is shown in Figure 12.6.

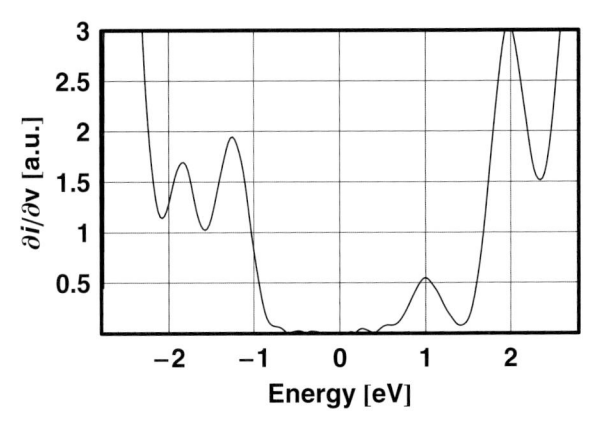

Fig. 12.6 The current derivative, $\partial i/\partial v$, as a function of v.

Finally, we calculate $\partial \ln i/\partial \ln v = (v/i)\partial i/\partial v$ as a function of v, which will be presented in the next section. The logarithmic derivative will be referred to as the LDOS of the surface of the sample, based on Eq. (12.9).

12.6
Comparison of STS Results

This section compares the experimental LDOS, obtained by a UHV scanning tunneling microscope, of C_{60} molecules adsorbed on Si(100)–(2 × 1), with energy levels of free C_{60} molecules. The results of the C_{60} energy-level calculations, based on the self-consistent-field discrete variational method, also agree with experimental results using X-ray photoemission spectra. Choosing the vacuum level as the zero of the energy scale, the calculations show that the highest occupied level, h_u, is at -6.04 eV, below which are two closely spaced levels, g_g and h_g, at -7.22 eV and -7.34 eV, respectively. The lowest unoccupied level, t_{1u}, is at -4.36 eV, above which is the level t_{1g} at -3.3 eV. Experimental scanning tunneling spectroscopic results of C_{60} molecules, chemisorbed on a Si(100)–(2 × 1) surface, yield peaks at 2.0 eV, 1.1 eV, -0.7 eV, and a broad peak at -2.3 eV. The "Fermi level" of a chemisorbed C_{60} molecule clearly shifts as it is adjusted to that of the substrate. We find that a shift of the energy levels of the free C_{60} molecule by 5.25 eV, given in Table 12.3, yields a good agreement with the experimentally-obtained STS data, shown in Figure 12.7. Here, the peaks at the bottom are the convoluted Fermi functions, under ambient conditions, at the appropriate free C_{60} energy levels. For this example, the linewidth of the observed spectra is larger than that which is determined by the temperature-dependent Fermi functions. Note the similarity between the calculated energy levels of the free C_{60} molecules and the STS curve.

Table 12.3 Free C_{60} energy-level calculations, based on the self-consistent-field discrete variational method.

Assignment energy (eV)
$t_{1g} = 1.97$
$t_{1u} = 0.94$
$h_u = -0.75$
$h_g = -1.92$
$g_g = -2.04$

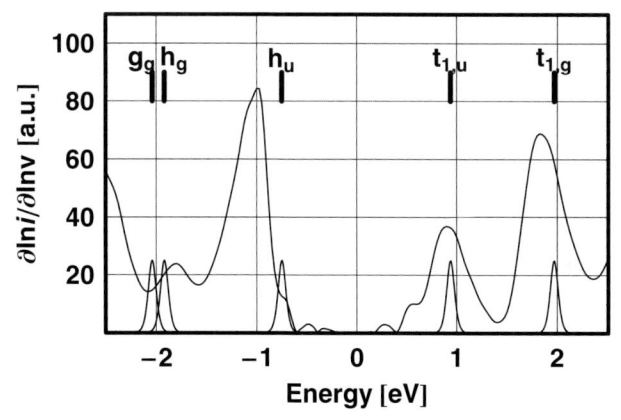

Fig. 12.7 The designation of the energy levels (top), the logarithmic derivative of the current (center), and the peaks of the convoluted Fermi functions (bottom).

▶ **Exercises for Chapter 12**

1. Find other examples in the literature where calculated energy levels fit experimental STS curves.
2. Consider 0-, 1-, 2-, and 3-dimensional systems and figure out the expected STS curves.

References

1. R. M. Feenstra, J. A. Stroscio, and A. P. Fein, "Tunneling spectroscopy of the Si(100)–2 × 1 surface," Surf. Sci. **181**, 295 (1987).
2. M. McEllistrem, G. Haase, D. Chen, and R. J. Hamers, "Electrostatic sample–tip interaction in the scanning tunneling microscope," Phys. Rev. Lett. **70**, 2471 (1992).
3. C. Julian Chen, *Introduction to Scanning Tunneling Micrscoopy*, Oxford University Press, New York, 1993.
4. X. Yao, T. G. Ruskell, R. K. Workman, D. Sarid, and D. Chen, "Scanning tunneling microscopy and spectroscopy of individual C_{60} molecules on Si(100)–2 × 1 surfaces," Surf. Sci. Lett. **366**, L743 (1996).

13
Coulomb Blockade

■ Highlights

1. Electrostatic energies and tunneling rates of a Coulomb staircase
2. Comparison of experiment and theory using an STM
3. Temperature, charge, and bias effects on the operation of an SET

Abstract

Nanometer-scale structures exhibiting Coulomb blockade and Coulomb staircase effects are in the frontier of electronics technology, being candidates to replace conventional transistors. A detailed theory of these effects and their experimental observation by scanning tunneling microscopy is given in Ref. [1] to Ref. [3]. This chapter presents an approximate model that takes into account temperature, charge, and bias effects, and reproduces the experimental results reported in these references.

13.1 Introduction

A schematic diagram of a system exhibiting Coulomb blockade (CB) and Coulomb staircase (CS) effects is shown in Figure 13.1(a), and an electronic circuit used to model these effects is shown in Figure 13.1(b). Figure 13.1(a) depicts the apex of a scanning tunneling microscope tip, modeled as a sphere, that hovers on top of a conducting spherical nanostructure that is separated from the surface of a flat conducting sample. The nanostructure is small enough that adding or removing a single electron from it will change its electrostatic energy by an amount that is larger than its thermal energy. Such a structure is called a quantum dot (QD) because of the 3D quantum confinement of its electronic wave functions whose energy levels are discrete. In the following sections, however, we will assume that the distances separating these energy levels can be ignored because they are much smaller than the QD charging energy. Now let the STM tip and the sample be connected to terminals A and B of Figure 13.1, respectively, between which a dc bias, v_{AB}, is applied. The QD, however, is free to assume its own potential depending on the terminals' bias and its own charge. Suppose now that the tip–QD and QD–sample distances are small enough that the bias gives rise to tunneling of electrons among these three bodies. This current is controlled by tunneling resistances, denoted by R_1 and R_2 in Figure 13.1(b).

The electrons, as they enter and leave the QD, will raise and lower its potential relative to that of the terminals. Suppose that the capacitance of the QD is small enough that adding a single electron to it will raise its potential, v_{QD}, such that $v_{QD} > v_{AB}$. Since such a situation is energetically unfavorable, tunneling through the QD will be blocked. The resumption of the tunneling current, therefore, will require an increase of the bias to satisfy $v_{AB} > v_{QD}$. This effect, which has been observed experimentally, is called the Coulomb blockade. The same argument follows if two electrons reside in the quantum dot, in which case it is required that $v_{AB} > 2v_{QD}$. Repeating this argument to a larger number of electrons residing in the QD leads to a situation where the tunneling current, as a function of the applied bias, will resemble a staircase. This effect, called the Coulomb staircase, demonstrates that tunneling is blocked for as long as the applied bias is not large enough to overcome the Coulomb blockade associated with n electrons residing in the QD. It is also possible to externally influence the potential of the quantum dot by coupling it to an extra capacitor or resistor, and thus control the bias required to overcome the Coulomb blockade. Here one obtains the so-called single-electron transistor, or SET. As the following examples demonstrate, observation of the CB, CS, and SET effects requires that the system's capacitances be at most several atto Farad, and the tunneling resistances larger than several mega Ohm. The following sections present an approximate model of the Coulomb blockade and Coulomb staircase effects, using the parameters given in Ref. [1] to Ref. [3], that reproduces their reported experimental results. In addition, the effect of temperature on the Coulomb staircase effect is also modeled and demonstrated by an example. To fully understand the model presented in this chapter, however, the reader is advised to consult these references.

13.2
Capacitance

Figure 13.1 shows two capacitors, C_1 and C_2, representing the capacitance of the tip–QD and QD–sample geometries, respectively. Such a representation would be correct if the tip–QD–sample structure consisted of two parallel plane capacitors that share one electrode. In that case the tip–sample capacitance could be neglected because the electric fields belonging to one

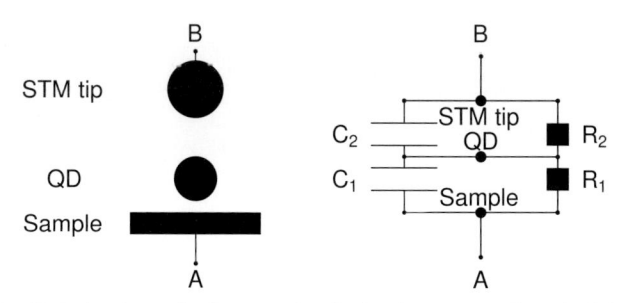

Fig. 13.1 A schematic diagram of a single-electron transistor (a: left-hand side) and its electric circuit representation (b: right-hand side).

capacitor do not penetrate into the other. For the tip–QD–sample geometry, however, one cannot obtain the capacitance of the whole system from that of each capacitor separately, since the electric fields in such a geometry surround both structures. Strictly speaking, one has to solve the three-body capacitance self-consistently to obtain the appropriate components of the capacitance associated with the tip–QD and QD–sample geometries. Nevertheless, we will approximate the total capacitance using the solutions to the isolated sphere–plane and sphere–sphere capacitances, as has been done extensively in the literature. The exact solution to the capacitance of a sphere–plane configuration is well known and can be coded concisely. For a sphere–sphere configuration, however, the solution is quite lengthy and executing the code is time consuming. However, as will be shown in Chapter 15, one can obtain a fairly accurate expression for the sphere–sphere capacitance that will execute much faster than the exact one.

13.2.1 Sphere–Plane Capacitance

It will be helpful to get a feel for the radii of the tip and the quantum dot, and the tip–dot and dot–sample separations required for enabling a Coulomb blockade effect. To that end, we present equations for sphere–plane and sphere–sphere capacitance together with several examples. An approximate solution to the sphere–plane capacitance, C_{sp}, is

$$C_{sp} = 2\pi\epsilon_0 r [\ln(s/r) + \ln(2) + 23/20], \qquad (13.1)$$

where r and s are the radius of the sphere and its separation from the surface of a sample, respectively. Figure 13.2 shows C_{sp} as a function of s for $r = 10$ (solid line), 20 (dashed line), and 30 nm (dotted line).

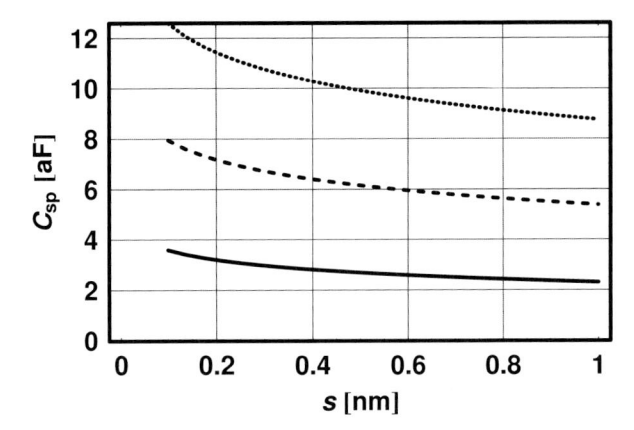

Fig. 13.2 The approximate sphere–plane capacitance, C_{sp}, as a function of their separation, s, for $r = 10$ (solid line), 20 (dashed line), and 30 nm (dotted line).

13.2.2
Sphere–Sphere Capacitance

An approximate solution to the sphere–sphere capacitance, C_{ss}, where r_1 and r_2 are the radii of the spheres and s is their separation, can be written in terms of C_{sp} as

$$C_{ss}(r_1, r_2, s) = \gamma C_{sp}(r, \gamma s), \tag{13.2}$$

where

$$\gamma = \frac{1}{1 + r_{min}/r_{max}}. \tag{13.3}$$

Here, r_{min} and r_{max} are the smaller and larger of the two radii, respectively, where r and s are the radius of the sphere and its separation from the surface of a sample, respectively. Figure 13.3 shows C_{ss} as a function of sphere–sphere separation, s, for $r_1 = r_2 = 10$ (solid line), $r_1 = r_2/2 = 10$ (dashed line), and $r_1 = r_2/3 = 10$ nm (dotted line).

13.3
Quantum Considerations

Let us now briefly review some of the parameters associated with the transfer of a single electron into a quantum dot.

(a) A quantized unit of resistance, R_Q, is

$$R_Q = \frac{\pi \hbar}{2e^2}. \tag{13.4}$$

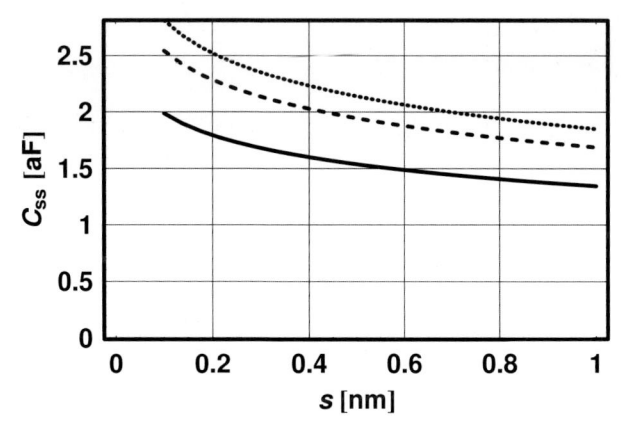

Fig. 13.3 The approximate capacitance between two spheres, C_{ss}, as a function of their separation, s, for $r_1 = r_2 = 10$ (solid line), $r_1 = r_2/2 = 10$ (dashed line), and $r_1 = r_2/3 = 10$ nm (dotted line).

(b) The thermal energy, E_{th}, at a given temperature, T, is

$$E_{th} = K_B T. \qquad (13.5)$$

(c) The change in the charging energy of a capacitor, due to a single electron entering (+) or leaving it (−), $\delta E_c^{\pm}$, is

$$\delta E_c^{\pm} = \frac{e^2}{2C}. \qquad (13.6)$$

(d) The time constant, τ, associated with charging and discharging of an RC circuit is

$$\tau = RC. \qquad (13.7)$$

(e) The quantum noise voltage, V_Q, of an RC circuit is

$$V_Q = \frac{\hbar}{eRC}. \qquad (13.8)$$

(f) The frequency associated with the number of tunneling electrons per second, f, is

$$f = \frac{i}{e}, \qquad (13.9)$$

which is related to the tunneling rates, γ.

13.4
Requirements and Approximations

The requirements for observing the Coulomb blockade and staircase effects are quite simple. First, the change in energy, $\delta E_c^\pm$, due to the addition or removal of a single electron from the quantum dot, must be large enough such that thermal excitations are below the CB threshold,

$$E_{th} \ll \delta E_c^\pm. \tag{13.10}$$

A second requirement is that the tunneling resistances, R_1 and R_2, be larger than the quantum resistance,

$$R_{12} \gg R_Q, \tag{13.11}$$

so that the quantum noise voltage is smaller than the CB threshold. As an example, consider the parameters given in Table 13.1 that yield the results shown in Table 13.2. We find that for the chosen parameters, both thermal fluctuations and quantum noise are indeed below $\delta E_c^\pm$, which makes it possible to observe the CB and CS effects in a system with such parameters.

Table 13.1 The parameters used as an example to assess the feasibility of observing a Coulomb effect.

$T = 300\,\text{K}$
$C = 1\,\text{aF}$
$R = 10\,\text{M}\Omega$
$i = 1\,\text{nA}$

Table 13.2 The parameters of a system that can exhibit the Coulomb effect.

$R_Q = 6.45319\,\text{k}\Omega$
$E_{th} = 25.8522\,\text{mV}$
$\delta E_c = 80.1089\,\text{mV}$
$\tau = 10\,\text{ps}$
$v_Q = 65.8211\,\mu\text{V}$
$f = 6.24151\,\text{GHz}$

13.5
Coulomb Blockade and Coulomb Staircase

A great simplification in the theory of the CB and CS effects occurs if the condition

$$R_2 \gg R_1 \tag{13.12}$$

is met, regardless of which of the two resistances is the larger one. This is usually satisfied anyway, since it is difficult to experimentally construct two identical STM junctions. This condition means that the junction with the highest resistance dictates the rate at which electrons tunnel in and out of the QD. Consequently, one can now use a simple analytic expression for the total tunneling rate, without having to involve quantum statistical considerations that are quite cumbersome.

13.5.1
Electrostatic Energy Due to the Charging of the Quantum Dot

Let the total charge on the quantum dot, Q, consist of n electrons plus an effective charge, Q_0, due to the difference in the work functions of the media comprising our system,

$$Q = ne - Q_0. \tag{13.13}$$

The electrostatic energy associated with increasing $(+)$ or decreasing $(-)$ the number of electrons in the quantum dot by 1, denoted by $\delta E_c^\pm$, is

$$\delta E_c^\pm = \frac{(Q \pm e)^2 - Q^2}{2(C_1 + C_2)}. \tag{13.14}$$

Note that here the two capacitors in Figure 13.1(b) appear to be in parallel rather than in series.

13.5.2
Electrostatic Energy Due to the Applied Bias

The electrostatic energies acquired by the capacitors C_1 and C_2 of Figure 13.1(b), ΔE_{v_1} and ΔE_{v_2}, respectively, due to the bias applied between terminals A and B, are

$$\delta E_{v_1}^\pm = \pm \frac{eC_2}{C_1 + C_2} v \tag{13.15}$$

and

$$\delta E_{v_2}^\pm = \pm \frac{eC_1}{C_1 + C_2} v. \tag{13.16}$$

13.5.3
Total Electrostatic Energy

The total electrostatic energy acquired by the capacitors C_1 and C_2, $\delta E_1^\pm$ and $\delta E_2^\pm$, respectively, is the sum of the charge and bias contributions. This energy is obtained by using Eq. (13.14) to Eq. (13.16), which yield

$$\delta E_1^\pm = \frac{e^2}{C_1 + C_2}\left[\frac{1}{2} \pm \left(n - \frac{Q_0}{e}\right) \pm \frac{C_2}{e}v\right] \quad (13.17)$$

and

$$\delta E_2^\pm = \frac{e^2}{C_1 + C_2}\left[\frac{1}{2} \pm \left(n - \frac{Q_0}{e}\right) \mp \frac{C_2}{e}v\right]. \quad (13.18)$$

Here, Q_0 accounts for a spurious charge residing in the QD.

13.6
Tunneling Rates

Let $\gamma_1^\pm$ and $\gamma_2^\pm$ denote the rates at which electrons tunnel through the tip–QD and QD–sample capacitors, C_1 and C_2, respectively, where the positive and negative signs refer to the direction of the tunneling. It can be shown that under the conditions stated by Eq. (13.10) to Eq. (13.12), the tunneling rates simplify to

$$\gamma_1^\pm = -\frac{\delta E_1^\pm}{e^2 R_1} \quad (13.19)$$

and

$$\gamma_2^\pm = -\frac{\delta E_2^\pm}{e^2 R_2}, \quad (13.20)$$

for $\delta E_i^\pm < 0$, and vanish otherwise.

13.7
Tunneling Current

The assumption that $R_2 \gg R_1$ leads to the result that the integer number of electrons residing in the quantum dot at any given time, n, satisfies

$$-C_2 v + Q_0 - \frac{e}{2} \leq n \leq -C_2 v + Q_0 + \frac{e}{2}. \quad (13.21)$$

It can be shown that the tunneling current, i, flowing between terminals A and B, is

$$i = i_0\left(\text{Floor}\left[\frac{-C_2 v + Q_0 + \frac{e}{2}}{e}\right], v\right), \quad (13.22)$$

where the brackets denote the integer part (Floor), and i_0 is given by

$$i_0 = e[\gamma_2^+(n,1,v) - \gamma_1^-(n,-1,v)]. \tag{13.23}$$

We can now choose the six parameters i_{max}, Q_0, C_1, C_2, R_1, and R_2, and plot the tunneling current as a function of the applied bias. For a proper choice of parameters, and at a low enough temperature, the resultant $i(v)$ curves should exhibit the Coulomb blockade and Coulomb staircase effects.

13.8 Examples

It is instructive to use examples from the literature where the theory presented in this chapter has been applied and compared to experiment. The following five such examples are taken from Ref. [1] and Ref. [2], reproducing their results. Note that in all the examples, $R_2 \gg R_1$, while the capacitances are comparable in magnitude. The following calculates the maximum voltage, v_{max}, n_1, n_2, and n_{max}, all of which are used to generate the plots presented in the following five examples.

$$v_{max} = i_{max}(R_1 + R_2), \tag{13.24}$$
$$n_1 = \text{Floor}[(-C_2 v_{max} + Q_0 + e/2)/e], \tag{13.25}$$
$$n_2 = \text{Floor}[(-C_2 v_{max} + Q_0 - e/2)/e], \tag{13.26}$$

and

$$n_{max} = \max\{n_1, n_2\}. \tag{13.27}$$

13.8.1
Example 1

This example uses the parameters given in Table 13.3 that generate the parameters given in Table 13.4.

Table 13.3 The parameters given in the first example.

i_{max}	= 1200 pA
Q_0	= −0.096 e
C_1	= 13.6 aF
R_1	= 0.3 MΩ
C_2	= 4.05 aF
R_2	= 29.3 MΩ

Table 13.4 The parameters calculated in the first example.

$$n_{max} = 1\,e$$
$$E_{th} = 0.361\,93\,\text{mV}$$
$$\delta E_c = 4.538\,75\,\text{mV}$$
$$\tau = 0.118\,665\,\text{ns}$$
$$v_Q = 0.161\,326\,\text{mV}$$

Figure 13.4 shows the Coulomb staircase generated in the first example that demonstrates two abrupt Coulomb staircases separated by a narrow Coulomb blockade, and a background of $i(v)$ slopes.

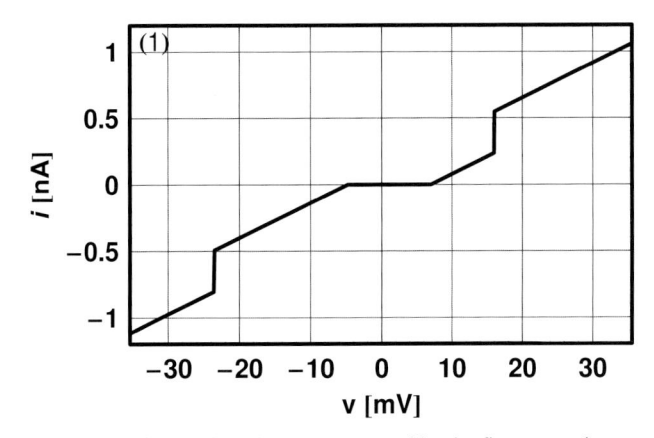

Fig. 13.4 The Coulomb staircase generated by the first example.

13.8.2
Example 2

This example uses the parameters given in Table 13.5 that generate the parameters given in Table 13.6.

Table 13.5 The parameters given in the second example.

$$i_{max} = 750\,\text{pA}$$
$$Q_0 = -0.005\,e$$
$$C_1 = 1.64\,\text{aF}$$
$$R_1 = 2\,\text{M}\Omega$$
$$C_2 = 3.28\,\text{aF}$$
$$R_2 = 39.2\,\text{M}\Omega$$

Table 13.6 The parameters calculated in the second example.

$$n_{max} = 1\,e$$
$$E_{th} = 0.361\,93\,\text{mV}$$
$$\delta E_c = 16.2823\,\text{mV}$$
$$\tau = 0.128\,576\,\text{ns}$$
$$v_Q = 0.200\,674\,\text{mV}$$

Figure 13.5 shows the Coulomb staircase generated in the second example that demonstrates two abrupt Coulomb staircases separated by a wide Coulomb blockade and a background of $i(v)$ slopes.

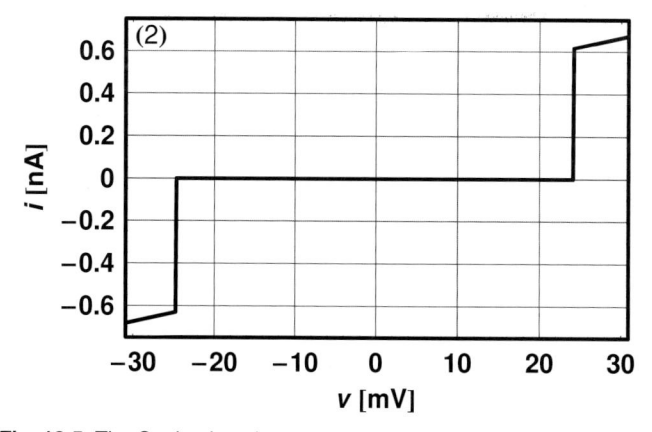

Fig. 13.5 The Coulomb staircase generated by the second example.

13.8.3
Example 3

This example uses the parameters given in Table 13.7 that generate the parameters given in Table 13.8.

Table 13.7 The parameters given in the third example.

$$i_{max} = 10\,900\,\text{pA}$$
$$Q_0 = -0.119\,e$$
$$C_1 = 0.938\,\text{aF}$$
$$R_1 = 1.7\,\text{M}\Omega$$
$$C_2 = 0.734\,\text{aF}$$
$$R_2 = 16.6\,\text{M}\Omega$$

Table 13.8 The parameters calculated in the third example.

$$n_{max} = 1\,e$$
$$E_{th} = 0.361\,93\,mV$$
$$\delta E_c = 47.912\,mV$$
$$\tau = 0.012\,184\,4\,ns$$
$$v_Q = 0.412\,775\,mV$$

Figure 13.6 shows the Coulomb staircase generated in the third example that demonstrates two abrupt Coulomb staircases separated by a Coulomb blockade, and a background of $i(v)$ slopes.

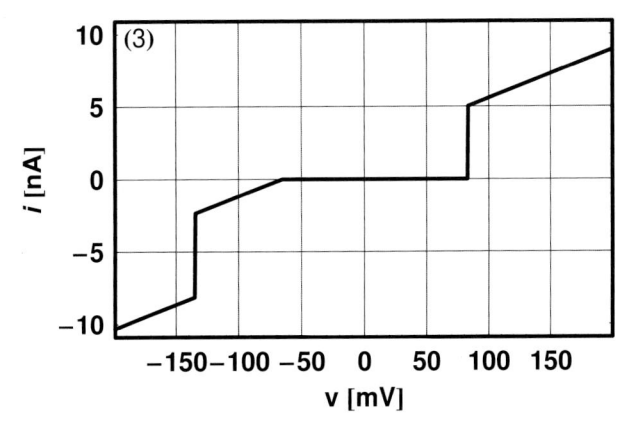

Fig. 13.6 The Coulomb staircase generated by the third example.

13.8.4
Example 4

This example uses the parameters given in Table 13.9 that generate the parameters given in Table 13.10.

Table 13.9 The parameters given in the fourth example.

$$i_{max} = 550\,pA$$
$$Q_0 = 0\,e$$
$$C_1 = 2\,aF$$
$$R_1 = 34.9\,M\Omega$$
$$C_2 = 4.14\,aF$$
$$R_2 = 132\,M\Omega$$

Table 13.10 The parameters calculated in the fourth example.

$n_{max} = 2e$
$E_{th} = 0.361\,93\,\text{mV}$
$\delta E_c = 13.047\,\text{mV}$
$\tau = 0.546\,48\,\text{ns}$
$v_Q = 0.009\,429\,95\,\text{mV}$

Figure 13.7 shows the Coulomb staircase generated in the fourth example that demonstrates four abrupt Coulomb staircases separated by a narrow Coulomb blockade, and a background of $i(v)$ slopes.

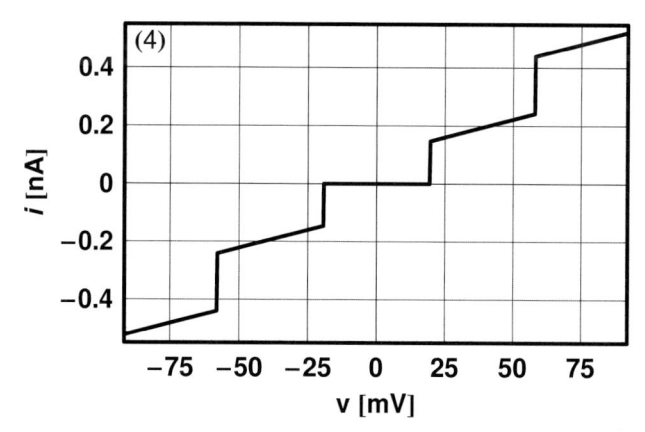

Fig. 13.7 The Coulomb staircase generated by the fourth example.

13.8.5
Example 5

This example uses the parameters given in Table 13.11 that generate the parameters given in Table 13.12.

Table 13.11 The parameters given in the fifth example.

$i_{max} = 65\,\text{pA}$
$Q_0 = 0\,e$
$C_1 = 1\,\text{aF}$
$R_1 = 1\,\text{M}\Omega$
$C_2 = 100\,\text{aF}$
$R_2 = 100\,\text{M}\Omega$

Table 13.12 The parameters calculated in the fifth example.

$$n_{max} = 4e$$
$$E_{th} = 0.361\,93\,\text{mV}$$
$$\delta E_c = 0.793\,157\,\text{mV}$$
$$\tau = 10\,\text{ns}$$
$$v_Q = 0.658\,211\,\text{mV}$$

Figure 13.8 shows the Coulomb staircase generated in the fifth example that demonstrates eight abrupt Coulomb staircases separated by narrow Coulomb blockades.

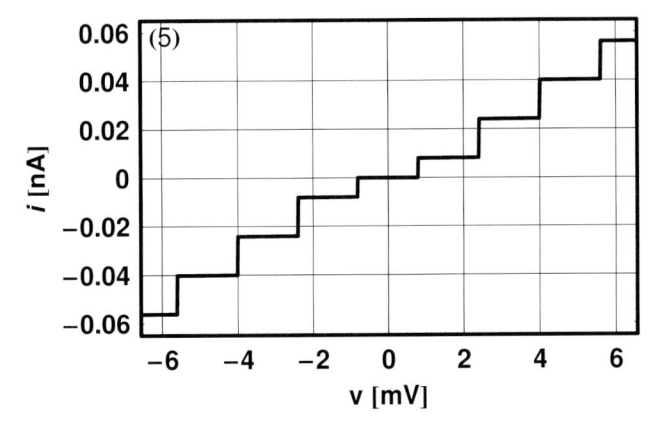

Fig. 13.8 The Coulomb staircase generated by the fifth example.

Note that these five examples are in good agreement with the experimental results obtained in Ref. [1] and Ref. [2].

13.9
Temperature Effects

13.9.1
Parameters

The theory presented in the last sections can be broadened to encompass the effect of temperature, T, and a possible external resistor, R_x, capacitor, C_x, chemical potential difference among the bodies of the SET, v_x, and spurious charge residing in the QD as described in Ref. [3]. Table 13.13 gives typical parameters chosen to demonstrate the effect of all these parameters on the operation of an SET.

Table 13.13 The parameters chosen for the examples demonstrating temperature effects.

$T = 2\,\text{K}$
$Q_0 = 0.3\,e$
$C_1 = 0.1\,\text{aF}$
$R_1 = 1\,\text{M}\Omega$
$R_2 = 100\,\text{M}\Omega$
$C_2 = 100\,\text{aF}$
$C_x = 1\,\text{aF}$
$v_x = 20\,\text{mV}$

13.9.2 Very High Temperature Operation

Let i_{linear} denote the current flowing through the sum of the two resistors of Figure 13.1(b) due to a bias v.

$$i_{\text{linear}} = v/(R_1 + R_2). \tag{13.28}$$

Figure 13.9, which shows the current as a function of the bias, will be used to demonstrate that the SET, at elevated temperatures, behaves as a linear circuit.

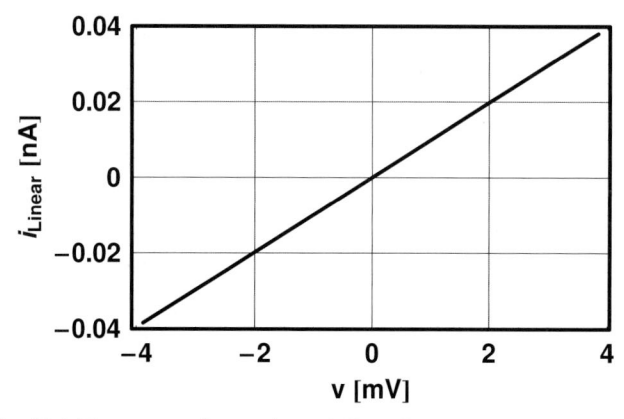

Fig. 13.9 The current, i_{Linear}, through $R_1 + R_2$ as a function of the bias, v.

13.9.3 Very Low Temperature Operation

At a very low temperature, T_0, the change in the charging energy of a capacitor due to a single electron entering (+) or leaving (−) it, E_c, is

$$E_c = \frac{e^2}{2(C_1 + C_2 + C_x)}. \tag{13.29}$$

The total electrostatic energy, $\delta E_1^\pm$ and $\delta E_2^\pm$, acquired by the two capacitors, C_1 and C_2, respectively, is the sum of the charge and bias contributions. This energy is obtained by using Eq. (13.14) to Eq. (13.16), which yield

$$\delta E_1^\pm(n,v) = \frac{e^2 \left[\frac{1}{2} \pm \left(n + \frac{Q_0 - v_x C_x}{e}\right) \mp \frac{C_2}{e} v\right]}{C_1 + C_2 + C_x} \tag{13.30}$$

and

$$\delta E_2^\pm(n,v) = \frac{e^2 \left[\frac{1}{2} \mp \left(n + \frac{Q_0 - v_x C_x}{e}\right) \mp \frac{(C_1 + C_x)}{e} v\right]}{C_1 + C_2 + C_x}. \tag{13.31}$$

Let $r_{0,1}^\pm$ and $r_{0,2}^\pm$ denote the rates at which electrons tunnel through the tip–QD and QD–sample capacitors, C_1 and C_2, respectively. Here the positive and negative signs refer to the direction of the tunneling. It can be shown that under the conditions stated by Eq. (13.10) to Eq. (13.12), the tunneling rates simplify to

$$r_{0,1}^\pm(n,v) = \frac{-\delta E_1^\pm(n,v)}{e^2 R_1} \tag{13.32}$$

and

$$r_{0,2}^\pm(n,v) = \frac{-\delta E_2^\pm(n,v)}{e^2 R_2}, \tag{13.33}$$

both of which are set to zero for positive $\delta E_i^\pm$. It can also be shown that the tunneling current, i_{T_0}, flowing between terminals A and B, is given by

$$i_{T_0}(n,v) = i_{0,3}(n,v) \left[\text{Floor}\left(\frac{C_2 v - Q_0 + C_x v_x + \frac{e}{2}}{e}\right), v\right], \tag{13.34}$$

where

$$i_{0,3}(n,v) = e[r_{0,2}^\pm(n,v) - r_{0,2}^\mp(n,v)]. \tag{13.35}$$

Figure 13.10 shows the current, i_{T_0}, as a function of bias, v, for the low temperature case. The operation of the SET here demonstrates a number of Coulomb staircases separated by Coulomb blockades.

13.9.4
Finite Temperature Operation

For a finite temperature, we have to take into account the Bose–Einstein occupation number, and the tunneling rates, Eq. (13.32) and Eq. (13.33), become

13.9 Temperature Effects

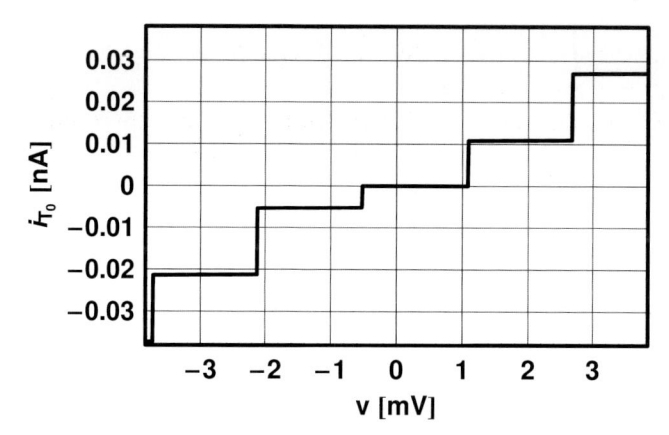

Fig. 13.10 A low-temperature SET demonstrates a number of Coulomb staircases separated by Coulomb blockades.

$$r_1^\pm(n,v) = -\frac{\delta E_1^\pm(n,v)}{e^2 R_1 \left(1 - e^{\frac{\delta E_1^\pm(n,v)}{k_B T}}\right)} \tag{13.36}$$

and

$$r_2^\pm(n,v) = -\frac{\delta E_2^\pm(n,v)}{e^2 R_2 \left(1 - e^{\frac{\delta E_1^\pm(n,v,Q_0)}{k_B T}}\right)}. \tag{13.37}$$

The temperature-dependent statistical function, ϱ, which describes the number of electrons for a given bias is

$$\varrho(n,v) = \frac{\left[\prod_{i=-n_{max}}^{n-1} x(i,v)\right]\left[\prod_{i=n+1}^{n_{max}} y(i,v)\right]}{\sum_{j=-n_{max}}^{n_{max}} \left\{\left[\prod_{i=-n_{max}}^{j-1} x(i,v)\right]\left[\prod_{i=j+1}^{n_{max}} y(i,v)\right]\right\}}. \tag{13.38}$$

Here,

$$x(n,v) = r_1^\pm(n,v) + r_2^\mp(n,v) \tag{13.39}$$

and

$$y(n,v) = r_1^\mp(n,v) + r_2^\pm(n,v). \tag{13.40}$$

Figure 13.11 depicts the statistical function, ϱ, as a function of the number of electrons, n, for a particular bias, v_{test}, as an example. The figure shows that at this bias the only occupied state is for $n = 2$.

Finally, the tunneling current, i, is given by

$$i(v) = \sum_{n=-n_{max}}^{n_{max}} e[r_2^\pm(n,v) - r_2^\mp(n,v)] \cdot \varrho(n,v) \tag{13.41}$$

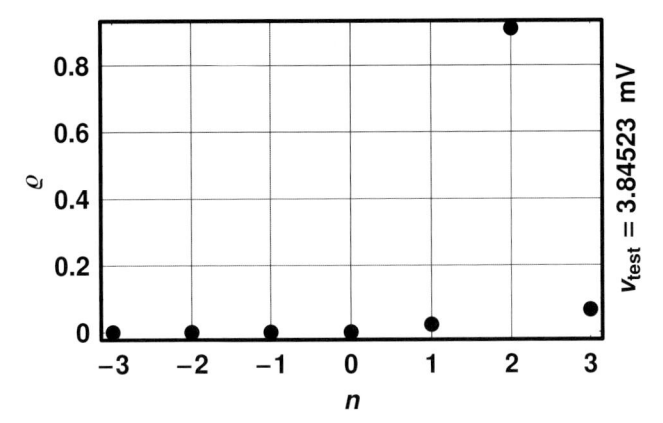

Fig. 13.11 The statistical function, ϱ, as a function of the number of electrons, n, for a bias v_{test}.

Figure 13.12 shows the tunneling current, i, for very low (straight line), finite (dots), and very high temperatures (staircase) as a function of bias, v. Note that the dots will approach the straight line as the temperature is elevated.

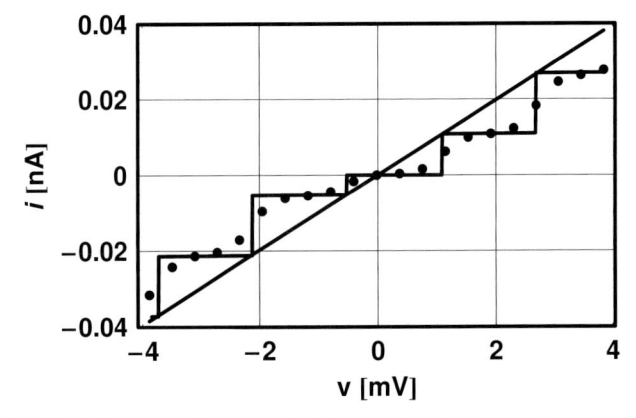

Fig. 13.12 The tunneling current, i, for very low (straight line), finite (dots), and very high temperatures (staircase) as a function of v.

▶ **Exercises for Chapter 13**

1. Check whether the model reproduces the experimental results cited in the references.

2. Choose your own parameters for R_j and C_j, $i = 1, 2$, and for Q_0, and T, and check whether the temperature and quantum noise are small enough to makes it possible to observe the CB and CS phenomena.
3. Extend the modeling so that it can handle a finite temperature.
4. Model the action of a single-electron transistor by controlling Q_0.

References

[1] A. E. Hanna and M. Tinkham, "Variation of the Coulomb staircase in a two-junction system by fractional electron charge," Phys. Rev. B **44**, 5919 (1991).

[2] M. Amman, R. Wilkins, E. Ben-Jacob, P. D. Maker, and R. C. Jaklevic, "Analytic solution for the current–voltage characteristic of two mesoscopic tunnel junctions coupled in series," Phys. Rev. B **43**, 1146 (1991).

[3] D. V. Averin and K. K. Likharev, "Single electronics: a correlated transfer of single electrons and Cooper pairs in systems of small tunnel junctions," in *Mesoscopic Phenomena in Solids*, B. L. Altshuler, P. A. Lee, and R. A. Webb, Eds., North-Holland, New York, 1991.

14
Density of States

■ **Highlights**

1. Density of states of large bodies for fractional dimensions
2. Density of states of quantum wells and quantum wires
3. Density of states of cubical and spherical quantum dots
4. Electron interband optical transitions near van Hove critical points

Abstract

The density of electronic states of solid bodies is a concept that is frequently encountered in the field of scanning probe microscopy in which structures on the nanometer scale are probed. The number of electronic states and their density is given for arbitrary dimension. The only equations we use are the volume of a sphere in a continuous dimension, and the energy of an electron in the parabolic approximation. *Mathematica* uses these two equations to generate the density of states of large bodies in three dimensions, quantum wells, quantum wires, and finds cubical and spherical quantum dots. Electron interband optical transitions near van Hove critical points are treated using fractional dimensions.

14.1
Introduction

The electronic density of states (DOS) of a solid body, ρ, plays a major role in solid-state physics, since it controls the possible interactions the body can have with its environment. In particular, for scanning tunneling microscopy applications, tunneling of electrons into or out of a body requires that it have available empty or filled states, respectively. For a body whose size in the $\hat{x}$, $\hat{y}$, and $\hat{z}$ axes is large relative to an atomic diameter, the density of states is a smooth function of the energy. However, if the geometry of the body is such that along at least one axis its size is only several atomic units, the density of states as a function of energy exhibits a particular structure. The DOS enters explicitly into the equations that govern the tunneling of electrons into or out of a given body. It is therefore important to calculate it for typical geometries encountered in the field of scanning tunneling microscopy. This chapter opens with a treatment of a volume in a d-dimensional space, where d is a real, positive number. The results are then applied to the calculation of the DOS of bodies in one, two, and three dimensions. Next, the DOS of a quantum well and a quantum wire, which are three-dimensional bodies confined along one and two axes, respectively, are discussed. Finally, the case of a quantum dot, which is a three-dimensional body confined along all the three axes and has a cubical or spherical geometry, is treated. The results presented here, which take into account spin degeneracy, are scaled to a volume $V = L^d$, where d is the dimensionality of the body and $L = 1$ nm. Also treated are electron interband optical transitions near van Hove critical points using fractional dimensions. Schematic diagrams of (a) three-dimensional, (b) two-dimensional, (c) one-dimensional, and (d) zero-dimensional bodies are shown in Figure 14.1.

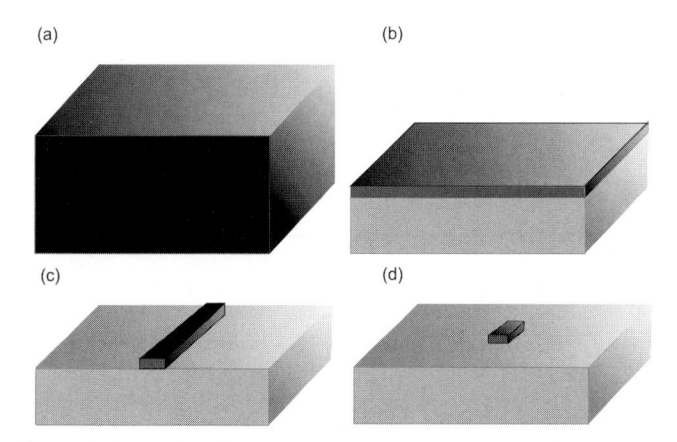

Fig. 14.1 Schematic diagrams of (a) three-dimensional, (b) two-dimensional, (c) one-dimensional and (d) zero-dimensional bodies.

14.2
Sphere in Arbitrary Dimensions

The volume, V_s, of a sphere with radius r in a d-dimensional space is

$$V_s = \frac{r^d \pi^{d/2}}{\Gamma[d/2 + 1]}. \tag{14.1}$$

Here, Γ is the Gamma function and d a real, positive number, not necessarily an integer. Equation (14.1) agrees with the well-known results for 0-, 1-, 2-, and 3-dimensional spaces, given in Table 14.1. Figure 14.2 shows an example of the volume of two spheres, one with a radius $r = 1.1$ (solid line) and the other with a radius $r = 1.2$ (dashed line), as a function of the continuous dimension d. We see that on increasing r, the volume of the spheres reaches a maximum at a larger value of d, and that the volume equals unity for $d = 0$. Driven by curiosity, we figure out the dimension, d_{max}, at which the volume obtains its maximum value for a given radius r. For example, for $r = 1.1$, we get $d_{max} = 6.5809$.

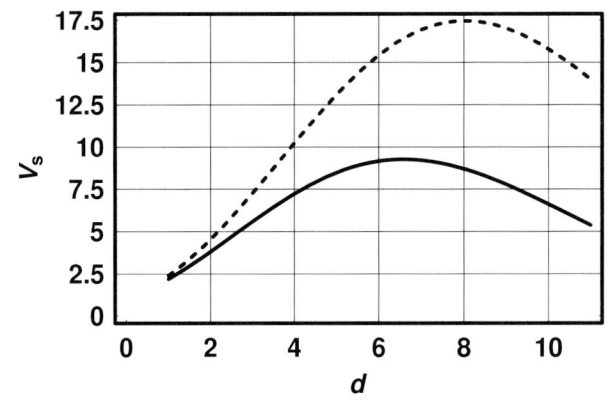

Fig. 14.2 The volume of two spheres, V_s, one with a radius $r = 1.1$ (solid line) and the other with a radius $r = 1.2$ (dashed line) as a function of the continuous dimension d.

Let A_s be the area of a sphere with radius r in a d-dimensional space, where d is a real, positive number. The area is obtained from the volume by taking its derivative with respect to r. The solid angle, θ_s, subtended by this sphere, is obtained by dividing the area of the sphere by r^{d-1}. Table 14.2 is a summary of the results obtained up to now for 1-, 2-, and 3-dimensions.

Table 14.1 The volume, V_s, area, A_s, and solid angle, θ_s, for a sphere in 1-, 2-, and 3-dimensions.

	$d = 1$	$d = 2$	$d = 3$
V_s	$2r$	πr^2	$\dfrac{4\pi r^3}{3}$
A_s	2	$2\pi r$	$4\pi r^2$
θ_s	2	2π	4π

14.3
Density of States in Arbitrary Dimensions

Using Eq. (14.1), we will now calculate the total number and density of states as a function of the energy of electrons for a body in a d-dimensional space. We will find that for integer dimensions, the results agree with the well-known expressions obtained in solid-state physics. Finally, we will plot the density of states as a function of a continuous dimension.

The volume in a d-dimensional k-space, V_k, is obtained by replacing r with k in Eq. (14.1), which yields

$$V_k = \frac{k^d \pi^{d/2}}{\Gamma[d/2+1]}. \tag{14.2}$$

The energy of an electron in the parabolic approximation, $\mathcal{E}$, is

$$\mathcal{E} = \frac{\hbar^2 k^2}{2m_e}. \tag{14.3}$$

Inverting Eq. (14.3) yields the energy-dependent k-vector, denoted by $k_\mathcal{E}$, in terms of the energy,

$$k_\mathcal{E} = \frac{\sqrt{2m_e \mathcal{E}}}{\hbar}. \tag{14.4}$$

The total number of k states, n_k, per unit volume in a d-dimensional space with dense k values, is

$$n_k = \frac{k^d}{2^{d/2-1} \pi^{d/2} \Gamma(d/2+1)}. \tag{14.5}$$

Here, a factor of 2 is included to account for the spin degeneracy, $1/(2\pi)^d$ is the number of k states per unit spatial volume, and V_k is the volume in k-space. Equation (14.5) gives the number of states, n_k, in terms of k. The number of states in terms of the energy, $n_{k,\mathcal{E}}$, is obtained by replacing k with $k_\mathcal{E}$

$$n_{k_\mathcal{E}} = \frac{\sqrt{2m_e \mathcal{E}}}{2^{d/2-1} \pi^{d/2} \hbar^d \Gamma(d/2+1)}. \tag{14.6}$$

To find the density of states, ρ, we take the derivative of $n_{k_\mathcal{E}}$ with respect to the energy $\mathcal{E}$

$$\rho = \frac{d(m_e \mathcal{E})^{d/2}}{(2\pi)^{d/2} \hbar^d \sqrt{\mathcal{E}} \Gamma(1+\frac{d}{2})}. \tag{14.7}$$

It is instructive to check the density of states per unit volume, obtained for a d-dimensional space, where $d = 1, 2$, and 3.

The density of states, ρ, as a function of energy, $\mathcal{E}$, for the $d = 1, 2$, and 3, using $L = 1$ nm, is shown in Figure 14.3.

Table 14.2 The density of states per unit volume, ρ, for $d = 1, 2,$ and 3.

	$d = 1$	$d = 2$	$d = 3$
ρ	$\dfrac{\sqrt{2}\sqrt{m_e}}{\pi\sqrt{\mathcal{E}}\hbar}$	$\dfrac{m_e}{\pi\hbar^2}$	$\dfrac{\sqrt{2}\sqrt{\mathcal{E}}m_e^{3/2}}{\pi^2\hbar^3}$

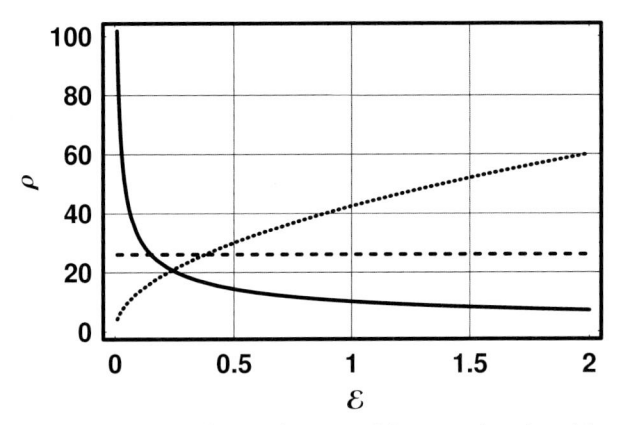

Fig. 14.3 The density of states (not to scale), ρ, as a function of the energy, $\mathcal{E}$, for $d = 1$ (solid line), 2 (dashed line), and 3 (dotted line).

The descending, flat, and ascending curves belong to the 1-, 2-, and 3-dimensional cases, respectively. The density of states as a function of both the energy, $\mathcal{E}$, and the continuous dimension, d, is shown in Figure 14.4.

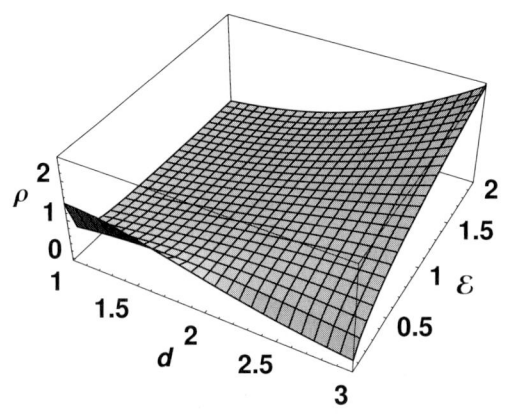

Fig. 14.4 The density of states, ρ, as a function of both the energy, $\mathcal{E}$, and the continuous dimension, d.

14.4
Density of States in Confined Structures

We will now treat three-dimensional bodies confined in one, two, and three dimensions, namely, quantum wells, quantum wires, and quantum dots. The first two bodies will have a finite density of states along the direction in which they are not confined, and discrete states in the direction in which they are confined. The quantum dots, on the other hand, will have only discrete states in all the three directions.

14.4.1
Quantum Wells

Let the quantum well have a thickness s along its confined direction, along which its energy states are discrete. The density of states along the unconfined directions, ρ_2, however, is two-dimensional. Now, the density of states of this body will have to equal that of a three-dimensional body at particular states where their k-vectors coincide.

$$\rho_2 = \frac{L}{s}\frac{m_e}{\pi\hbar^2}\,\text{Floor}\left[\frac{\sqrt{2m_e\mathcal{E}}}{\pi\hbar}s\right]. \tag{14.8}$$

Figure 14.5 shows the density of states of a quantum well and of a three-dimensional body. Note that the staircase structure, associated with the quantum well, contacts that of the three-dimensional body at discrete points.

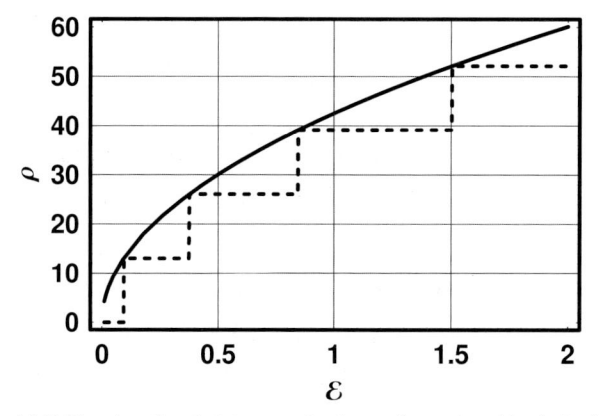

Fig. 14.5 The density of states, ρ, of a three-dimensional body (solid line) and a quantum well (dashed line) as a function of energy, $\mathcal{E}$.

14.4.2
Quantum Wires

Let the quantum wire have a thickness, s, in each of its confined directions. This thickness dictates that in these directions its energy states are discrete. The density of states along the unconfined direction, ρ_1, is one-dimensional. Now, the density of states of this body will have to equal that of a three-dimensional body at particular states where their k-vectors coincide. Using the function Floor$[x]$ that gives the greatest integer less than or equal to x, will accomplish this task, yielding

$$\rho_1 = \frac{L^2}{s^2} \frac{\sqrt{2m_e/\mathcal{E}}}{\pi \hbar} \text{Floor}\left[\frac{m_e \mathcal{E}}{\pi \hbar^2} s^2\right]. \tag{14.9}$$

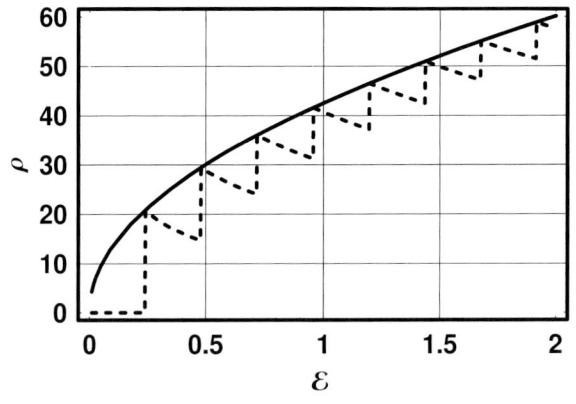

Figure 14.6 shows the density of states of a three-dimensional body (solid line) and a quantum wire (dashed line) as a function of energy, $\mathcal{E}$. Note that the structure of the quantum wire contacts that of the three-dimensional body at discrete points.

Fig. 14.6 The density of states of a three-dimensional body (solid line) and a quantum wire (dashed line) as a function of energy, $\mathcal{E}$.

14.4.3
Cubical Quantum Dots

We will now consider a cubical quantum dot whose size, L_c, is small enough such that it can support only a discrete, rather than a continuous, number of electronic states, $\mathcal{E}_n$. As the size of the quantum dot grows, the energy levels will start merging, eventually building up into a semicontinuous density of states. The energy levels for this structure, enumerated by n, are obtained by considering the energy of an electron in the parabolic approximation, replacing the k-vector by $n\pi/L_c$, and multiplying the energy by 3 to account for the three degrees of freedom in three-dimensional space,

$$\mathcal{E}_n = \frac{3\pi^2 \hbar^2}{2m_e L_c^2} n^2. \tag{14.10}$$

Figure 14.7 shows the energy levels of an electron in the cubical quantum dot, $\mathcal{E}_n$, for $n = 1 - 3$. The lowest energy level is shown by the solid line, with ascending curves for higher energies. Note that as the size L_c increases, the energy levels develop into a semicontinuous density of states.

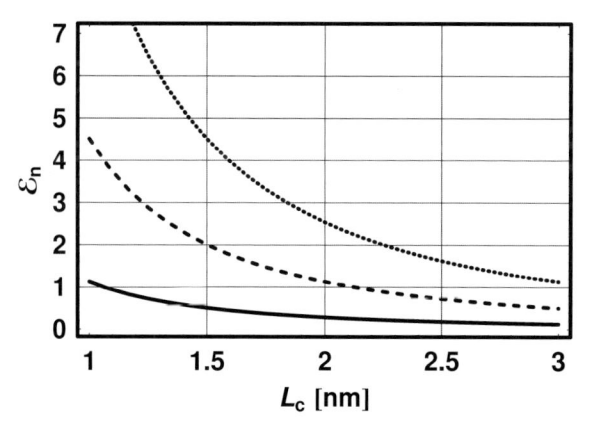

Fig. 14.7 The energy levels, $\mathcal{E}_n$, of a cubical quantum dot as a function of L_c.

14.4.4
Spherical Quantum Dots

We will now consider a spherical quantum dot whose radius, R, is small enough such that it can support only a discrete, rather than a continuous, number of electronic states ([1], p. 337). The first six energy levels of the quantum dot, $\mathcal{E}_n$, as a function of its radius will now be calculated. As with the cubical quantum dot, here too we find that as the radius of the quantum dot increases, the energy levels merge into a semi continuous density of states. The solution to Schrodinger's equation for a single electron,

$$-\frac{\hbar^2}{2m_e} \nabla^2 \psi_s(r) = E_s \psi_s(r), \tag{14.11}$$

subject to the spherically symmetric boundary condition

$$\psi_s(R) = 0, \tag{14.12}$$

where $|r| < R$, is given by

$$\psi_s = \frac{2^{1/2}}{R^{3/2}} \frac{j_1(\alpha_{n,\ell} r/R)}{j_{1+\ell}(\alpha_{n,\ell})} Y_{\ell,m}(\Omega). \tag{14.13}$$

Here, j_1 is a spherical Bessel function of order 1 and $Y_{\ell,m}$ is a spherical harmonic. For the boundary conditions of ψ_s to be satisfied, it is required that

$$j_1(\alpha_{n,\ell}) = 0. \tag{14.14}$$

The first six zeros of $j_1(\alpha_{n,\ell})$ are given in Table 14.3 for a particular choice of (n, ℓ).

Table 14.3 The first six zeros of $j_1(\alpha_{n,\ell})$ for a particular choice of (n, ℓ).

(0,1)	(0,2)	(0,3)	(1,1)	(1,2)	(1,3)
3.14159	6.28319	9.42478	4.49341	7.72525	10.9041

Figure 14.8 shows the energy levels of an electron in the spherical quantum dot, $\mathcal{E}_{n,\ell}$, as a function of its radius, R, using the six functions listed in Table 14.3. The lowest energy level is shown by the solid line, with ascending curves referring to higher energies. Note that as the radius R increases, the energy levels develop into a semicontinuous density of states.

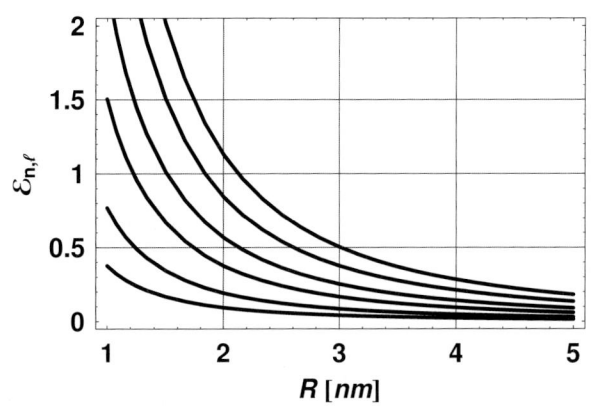

Fig. 14.8 The energy levels of a spherical quantum dot, $\mathcal{E}$, as a function of its radius, R, with ascending curves referring to higher energies.

To compare the energy levels of the spherical and cubical quantum dots, we consider a cube whose volume equals that of a sphere, namely $L_c^3 = (4/3)\pi R^3$. The ratio of the energies of their first level is found to be 0.866173, quite close to unity.

14.5
Interband Optical Transitions and Critical Points

Differentiation and integration to arbitrary order have been a topic of much interest for decades, and applications to many fields of science have been found very useful [3]. The common term for such operations, differintegration, implies that one can formulate an operator, $\mathcal{D}^q = d^q/dx^q$, whose parameter q determines whether it will differentiate or integrate a function $f(x)$. Moreover, q does not have to be an integer, but rather can assume fractional and even complex values. Among several definitions of such an operator is the one defined along the Riemann–Liouville integral, given by

$$\frac{d^q f(x)}{d(x-a)^q} = \frac{1}{\Gamma(-q)} \int_a^x \frac{f(y)}{(x-y)^{q+1}} dy, \qquad (14.15)$$

for $q < 0$, which can be extended to $q > 0$ by using

$$\frac{d^q f(x)}{d(x-a)^q} = \frac{d^n}{dx^n} \frac{1}{\Gamma(n-q)} \int_a^x \frac{f(y)}{(x-y)^{q-n+1}} dy, \qquad (14.16)$$

with $n > 0$. Equations (14.15) and (14.16) are mathematical operators whose validity can be justified by pure logical arguments. The notion of a fractional space, treated in the previous sections, is by far more complicated as it requires, for example, an explanation of the meaning of boundary conditions in such a space. These, and similar arguments, will not be dealt here.

In this section we give an example of how differintegration can be put to use in solving a physical problem. Consider the optical properties associated with electron interband transitions in an anisotropic solid in the vicinity of van Hove critical points. The Bloch electrons in this solid can be treated as if they were an isotropic electron gas residing in a fractional space of order α. The joint density of states here, ρ_e, is

$$\rho_e = \frac{2}{\Gamma(\alpha/2)} \left(\frac{m_{vc}}{2\pi\hbar^2}\right)^{\alpha/2} (\mathcal{E} - \mathcal{E}_g)^{d/2-1}, \qquad (14.17)$$

where $\mathcal{E}$ is the optical energy, $\mathcal{E}_g$ the band gap and m_{vc} the effective mass of the electron. The dielectric constant of the solid, $\epsilon(\hbar\omega)$, associated with allowed interband transitions near a critical point, in terms of $\rho_e(\mathcal{E})$, is

$$\epsilon(\hbar\omega) = \frac{C}{(\hbar\omega)^2} \int_{\rho_e(\mathcal{E})} \frac{\rho_e(\mathcal{E})}{\mathcal{E} - \hbar\omega - i\gamma} d\mathcal{E}, \qquad (14.18)$$

where ω is the optical radial frequency, C an electronic transition matrix element, and γ a broadening factor. Combining Eqs (14.17) and (14.18) yields

$$\epsilon(\hbar\omega) \propto \frac{i^{r-\alpha}}{(\hbar\omega)^2 \Gamma(\alpha/2)} \int^{\mathcal{E}_g} \frac{(\mathcal{E}_g - \mathcal{E})^{\alpha/2-1}}{\mathcal{E} - \hbar\omega - i\gamma} d\mathcal{E}, \qquad (14.19)$$

where r is the type of the critical point. Within the spirit of Eq. (14.15), we can write Eq. (14.19) around $\hbar\omega \sim \mathcal{E}_g$ as

$$\epsilon^{(k)}(\hbar\omega) \propto i^{r-\alpha} D^{-\alpha/2} \frac{1}{\mathcal{E}_g - \hbar\omega - i\gamma}, \qquad (14.20)$$

where the derivative is in respect to $\mathcal{E}_g$. Performing the differintegration yields

$$\epsilon(\hbar\omega) \propto i^{r-\alpha} \Gamma(2-\alpha/2) \int^{\mathcal{E}_g - \hbar\omega - i\gamma} t^{d/2-2}\, dt. \qquad (14.21)$$

Equation (14.21) gives the dielectric function, ϵ, for an α-dimensional solid. The integral of Eq. (14.21), for $r = 0$ and $\alpha = 0, 1, 2,$ and 3, is given in Table 14.4, where $x = (\hbar\omega - E_g)/\gamma$.

Table 14.4 ϵ as a function of $x = (\hbar\omega - E_g)/\gamma$, for $r = 0$ and $\alpha = 0, 1, 2,$ and 3.

α	ϵ
0	$-\dfrac{1}{i+x}$
1	$\dfrac{i\sqrt{\pi}}{\sqrt{i+x}}$
2	$-\ln(i+x)$
3	$2i\sqrt{\pi}\sqrt{i+x}$

Figure 14.9 shows $\Re(\epsilon)$ and $\Im(\epsilon)$ as a function of x for $\alpha = 0, 1, 2,$ and 3. To demonstrate the utility of Eq. (14.21), we also give ϵ as a function of $x = (\hbar\omega - E_g)/\gamma$, for $r = 0$ and $\alpha = 1.5$ and 2.5.

Table 14.5 ϵ as a function of $x = (\hbar\omega - E_g)/\gamma$, for $r = 0$ and $\alpha = 1.5$ and 2.5.

α	ϵ
1.5	$2.563\,\dfrac{1+i}{(i+x)^{0.25}}$
2.5	$2.5636\,\dfrac{1+i}{(i+x)^{0.25}}$

Figure 14.10 shows $\Re(\epsilon)$ and $\Im(\epsilon)$ as a function of x for $\alpha = 1.5$ and 2.5, relevant to a solid in a fractional dimension. There are many other cases where differintegration can help solve problems associated with spaces having fractional dimensions.

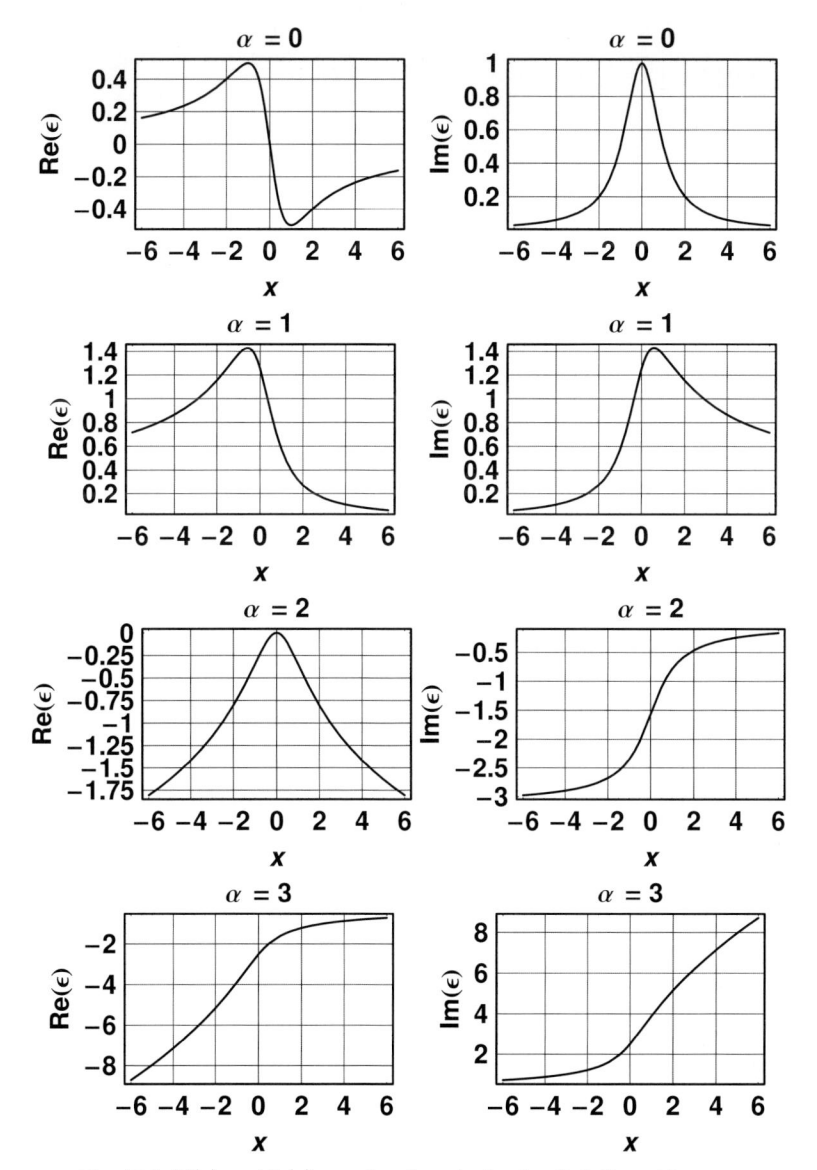

Fig. 14.9 $\Re(\epsilon)$ and $\Im(\epsilon)$ as a function of x for $d = 0, 1, 2$, and 3.

▶ **Exercises for Chapter 14**

1. Compare the first five energy levels of cubical and spherical quantum dots.
2. Calculate the density of states in continuous d-dimensions, taking into account the Fermi functions at a finite temperature.

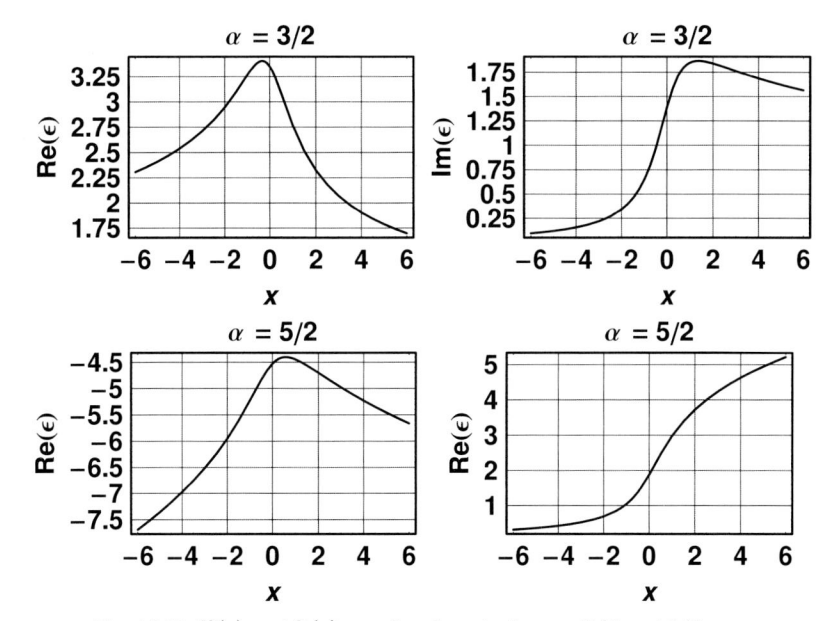

Fig. 14.10 $\Re(\epsilon)$ and $\Im(\epsilon)$ as a function of x for $\alpha = 3/2$ and $5/2$.

References

1. H. Haug and S. W. Koch, *Quantum Theory of the Optical and Electronic Properties of Semiconductors*, World Scientific, Singapore, 1990.
2. S. L. Chuang, *Physics of Optoelectronic Devices*, Wiley, New York, 1995.
3. K. B. Oldham and J. Spanier, *The Fractional Calculus: Theory and Applications of Differentiation and Integration to Arbitrary Order*, Academic Press, New York, 1974.
4. Xing-Fei He, "Fractional dimensionality and fractional derivative spectra of interband optical transitions," Phys. Rev. B **42**, 11751 (1990).

15
Electrostatics

■ Highlights

1. Exact and approximate sphere–plane and sphere–sphere capacitances
2. Electrostatic force between a conducting sphere and a conducting plane

Abstract

Scanning probe microscopy often requires solutions to the capacitance of a conducting sphere next to a conducting plane, and of two conducting spheres next to each other. We will obtain exact solutions to the capacitance of these bodies together with faster running, approximate solutions.

15.1
Introduction

It is often convenient to model the apex of a sharp conducting tip of a scanning probe microscope by a sphere, and the flat conducting sample by a plane. There are also cases where the tip scans over a conducting convex object that can also be modeled as a sphere. The electrostatic interaction between the tip and sample, which plays a major role in scanning probe microscopy in determining the tunneling current and electrostatic force, can then be modeled using sphere–plane and sphere–sphere geometries. We will use the method of images to obtain solutions to geometries consisting of a point charge and a plane, a point charge and a sphere, a sphere and a plane, and finally two spheres. We will obtain the electrostatic fields and equipotentials for all but the last case together with typical examples that treat nanometer-scale structures. We will also find exact solutions to the capacitance of sphere–plane and sphere–sphere geometries together with numerical examples. We can then study a particular case by using a replacement rule in the code that we can later modify if necessary. To speed up the computations, we will develop approximate solutions and test their range of validity. The fast-running approximate solutions will be sufficient because radii and distances associated with geometries encountered in scanning probe microscopy can only be crudely estimated from experimental results.

15.2
Isolated Point Charge

We will solve electrostatic problems by the method of images which uses a collection of isolated point charges to obtain exact solutions involving finite-size conducting bodies. We start therefore by treating the case of a point charge and use the results as building blocks for the solution of the more complicated cases. The potential, v_{point}, surrounding an isolated point charge, Q, located at the axes' origin, is given as a function of distance, r, by

$$v_{point} = \frac{1}{4\pi\epsilon_0} \frac{Q}{r}. \tag{15.1}$$

For simplicity of plotting, we will ignore the first factor on the right-hand side of Eq. (15.1) and assume that $Q = 1$. The field, E_{point}, surrounding the point charge, is obtained from the gradient of the potential by

$$E_{point} = -\nabla v_{point}. \tag{15.2}$$

The two-dimensional plots of the field lines (a) and equipotentials (b) of an isolated point charge are shown in Figure 15.1.

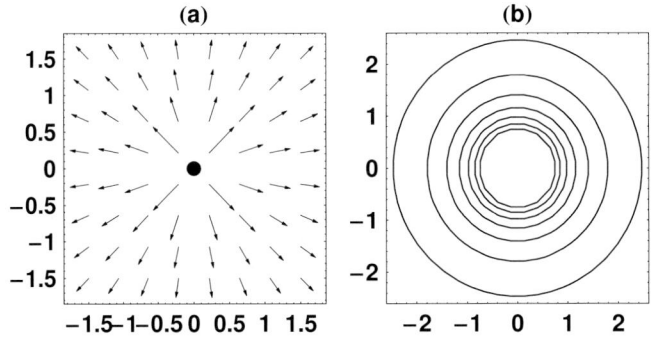

Fig. 15.1 The field lines (a) and equipotentials (b) of an isolated point charge.

15.3
Point Charge and Plane

We now place an infinite, grounded conducting plane at a distance d below a point charge, as shown in Figure 15.2(a), and use the method of images to solve for the field and potential surrounding these two bodies. The point charge will induce an image $Q_i = -Q$ at a position $s = -2d$, to satisfy the condition that the plane must be an equipotential. We can therefore replace the conducting plane by its image charge and solve for the field and potential using only the real and induced charges. The two-dimensional plots of the field lines (a) and equipotentials (b) of the point charge and plane are shown in Figure 15.2. In contrast to the previous case, the equipotentials and field lines now have only an axial symmetry in the three-dimensional space. Note that the field lines at the surface of the plane are at a right angle, and that the solutions to the field and potential are valid only outside the conducting body.

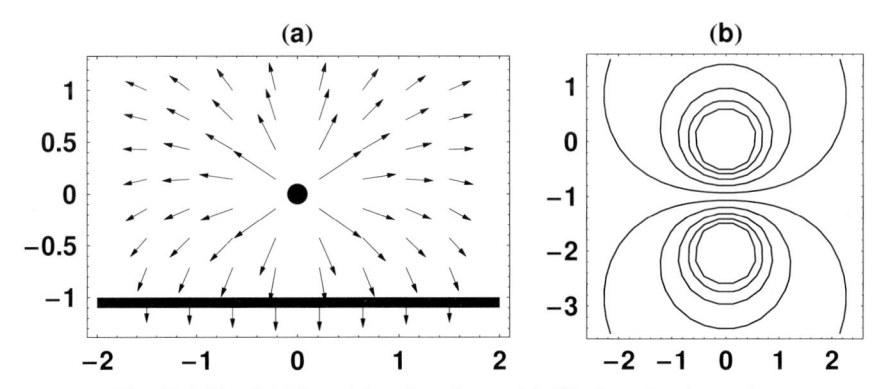

Fig. 15.2 The field lines (a) and equipotentials (b) of a point charge in the vicinity of a conducting plane.

15.4 Point Charge and Sphere

We now replace the conducting plane of Figure 15.2(a) with a conducting, grounded sphere centered at $-d$, as shown in Fig. 15.3(a), and use the method of images to solve for the field and potential surrounding these two bodies ([1], p. 150). The point charge will induce an image Q_i at a position s_i inside the sphere, to satisfy the condition that its surface must be an equipotential. We can therefore replace the conducting sphere by the image charge and solve for the field and potential using only the real and induced charges. Using geometrical considerations dictates that the position and magnitude of the image induced by the real charge, s_i and Q_i, be given by

$$s_i = d - \frac{R^2}{d} \tag{15.3}$$

and

$$Q_i = -Q\frac{R}{d}. \tag{15.4}$$

The two-dimensional plots of the field lines (a) and equipotentials (b) of a point charge and a conducting sphere are shown in Figure 15.3. Here, the potential produces a circle of radius R centered at $-d$, as expected, and the field lines are perpendicular to the surface of the sphere–sphere in a three-dimensional space. As noted before, the solution to the field and potential are valid only outside the conducting body.

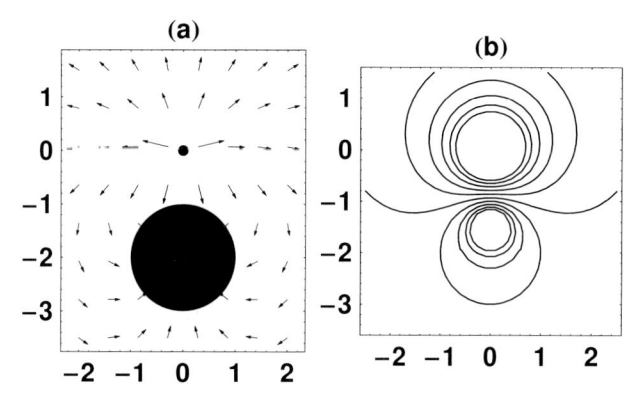

Fig. 15.3 The field lines (a) and equipotentials (b) of a point charge in the vicinity of a conducting sphere.

15.5
Isolated Sphere

The field and potential surrounding an isolated conducting sphere with radius R and charge Q are identical to those of a point charge for distances r such that $r > R$. However, inside the sphere, the potential is finite and constant, and the field vanishes. Now the capacitance of a conducting body is defined as the ratio of its charge to the potential it produces. Therefore, if the charge of the body is known, we can find the capacitance of the body by evaluating its potential. The self-capacitance of an isolated sphere, C_{sphere}, with radius R, is given by

$$C_{sphere} = 4\pi\epsilon_0 R. \tag{15.5}$$

As an example, note that a sphere with a radius of 1 nm has a capacitance of 0.111 aF. Such a capacitance is so small that adding to it a single electron will already change its potential by $v = 0.69$ V. This raises an interesting question, which is, what happens to the potential of such a sphere if we try to charge it through a tunneling process by applying a bias that is smaller than v. This important issue was discussed in detail in the chapter dealing with Coulomb blockade.

15.6
Sphere and Plane

Let us now place a conducting sphere with radius R at the axes' origin, and a conducting, grounded plane at $-d$, as shown in Figure 15.4(a). Our plan is to solve for the potential and field surrounding these two bodies, using the results obtained in the previous sections. Specifically, we use the solutions obtained with a point charge next to a plane and a point charge next to a sphere, and keep track of the magnitude and location of the real and induced charges. The potential generated by these charges, outside the conducting bodies, can then be calculated.

15.6.1
Position of Charges Inside the Sphere

The procedure starts by placing a unit charge at the axes' origin, which will make the surface of the sphere an equipotential. This charge will then induce an image having the same magnitude but opposite sign in the conducting half-space bounded above by the plane. The combination of these two will make the plane an equipotential, but the surface of the sphere will no longer be an equipotential. To improve the accuracy of the solution, so that both the surface of the sphere and the plane become equipotentials, we continue adding images of images iteratively, until the process converges to a close enough approximation for the actual potential of the sphere and plane. The position of the charges inside the sphere, up to three iterations, is shown in Table 15.1.

Table 15.1 The position of the charges inside a sphere positioned next to a conducting plane, up to three iterations.

$$\left\{0, \frac{R^2}{2d}, \frac{R^2}{2d - \frac{R^2}{2d}}\right\} \tag{15.6}$$

15.6.2
Magnitude of Charges Inside the Sphere

Table 15.2 gives the magnitude of the charges associated with their respective positions, presented in Table 15.1. Note that the magnitude of the real charge is unity and that the position of the image inside the sphere is at $R/2d$.

Table 15.2 The magnitude of the charges inside a sphere positioned next to a conducting plane, up to three iterations.

$$\left\{1, \frac{R}{2d}, \frac{R^2}{4d^2\left(1 - \frac{R^2}{4d^2}\right)}\right\} \tag{15.7}$$

15.6.3
Position of Charges Outside the Sphere

The position of the charges in the conducting half-space bounded above by the plane is given in Table 15.3. Note, for example, that the charge placed at the axes' origin induces its first image at a distance of $2d$ below the $\hat{x}$ axis.

Table 15.3 The position of the charges outside a sphere positioned next to a conducting plane, up to three iterations.

$$\left\{2d, 2d - \frac{R^2}{2d}, 2d - \frac{R^2}{2d - \frac{R^2}{2d}}\right\} \tag{15.8}$$

15.6.4
Magnitude of Charges Outside the Sphere

The magnitude of the charges associated with their respective positions, as shown in Table 15.3, is given in Table 15.4. Clearly the position and magnitude of the images given by the first two elements in the four lists reproduce the results obtained in the sections dealing with the point charge next to a plane and a point charge next to a sphere.

Table 15.4 The magnitude of the charges outside a sphere positioned next to a conducting plane, up to three iterations.

$$\left\{-1, -\frac{R}{2d}, -\frac{R^2}{4d^2\left(1-\frac{R^2}{4d^2}\right)}\right\} \quad (15.9)$$

15.6.5
Potential and Field

Figure 15.4 shows the field lines (a) and equipotentials (b) of a conducting sphere next to a conducting plane. Note that the equipotential at the surface of the plane is not a straight line because we have not taken enough iterations.

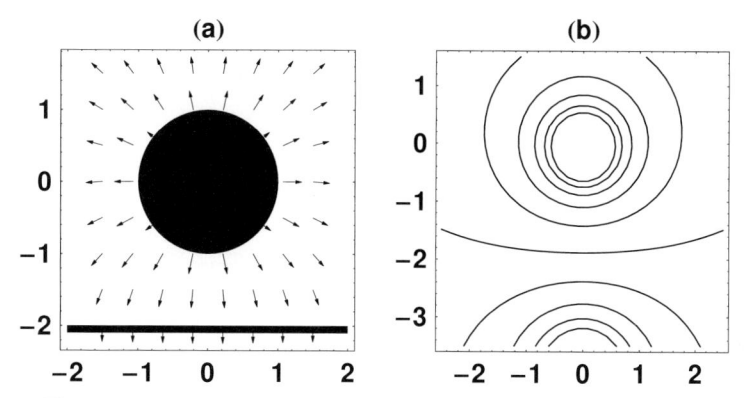

Fig. 15.4 The field lines (a) and equipotentials (b) of a conducting sphere next to a conducting plane.

15.6.6
Potential Along the Axis of Symmetry

Figure 15.5 shows the potential of a conducting sphere next to a conducting plane along the axis of symmetry for which $x = 0$.

15.7
Capacitance

The exact solution to the capacitance of a conducting sphere with radius R, C_{sp}, whose surface is separated from a conducting plane by a distance s, can be obtained by summing up the potential generated by all the charges, real and induced, as done before.

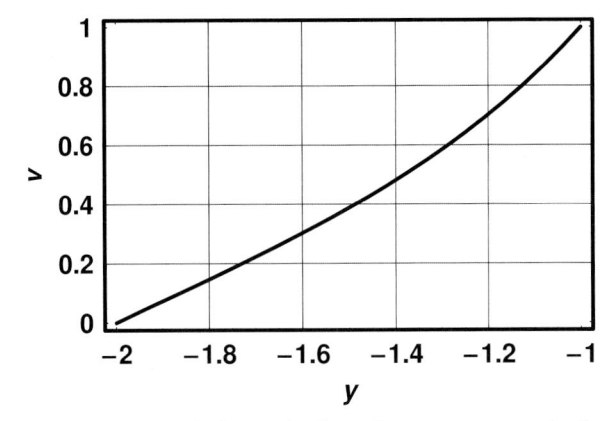

Fig. 15.5 The potential of a conducting sphere next to a conducting plane along the axis of symmetry for which $x = 0$.

15.7.1
Sphere–Plane Capacitance

The exact solution to the sphere–plane capacitance, C_{sp} (Figure 15.6), is

$$C_{sp} = 4\pi\epsilon_0 R \sinh\alpha \sum_{n=1}^{\infty} \frac{1}{\sinh(n\alpha)}, \qquad (15.10)$$

where

$$\alpha = \ln\left(\frac{s}{R} + \sqrt{\frac{s^2}{R^2} + \frac{2s}{R} + 1}\right). \qquad (15.11)$$

For $s < R$, one can approximate C_{sp} using

$$C_{sp} \sim 2\pi\epsilon_0 R[\ln(s/R) + \ln(2) + 23/20]. \qquad (15.12)$$

15.7.2
Example

15.8
Two Spheres

Consider now two conducting spheres with radii R_1 and R_2 whose surfaces are separated by a distance s. The exact solution to the potential and field of these two bodies, using the method of images, is quite lengthy and will not be given here. However, the code for their capacitance, $C_{ss}(R_1, R_2, s)$, is relatively short and will be used in the following section.

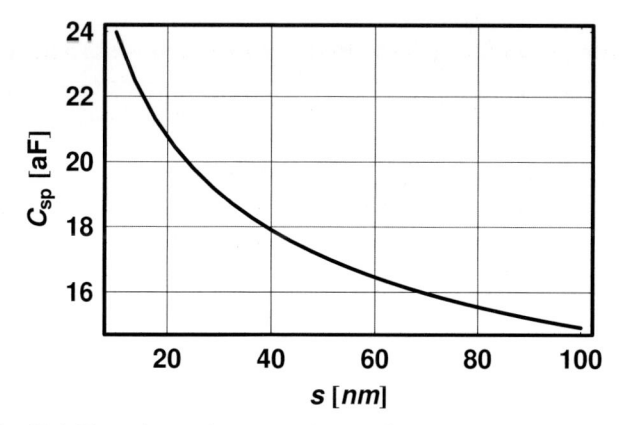

Fig. 15.6 The sphere–plane capacitance, C_{sp}, as a function of their separation, s.

15.8.1
Capacitance: Exact Solution

The capacitance of the two spheres is calculated by using all the building blocks obtained in the previous sections and will be given here without proof [2]. The solution is given in terms of an infinite series, and one has to choose the number of terms, n, in the summation that will yield a solution that is accurate enough. Usually, the range $10 < n < 20$ will be sufficient.

15.8.2
Capacitance: Approximate Solution

The approximate solution to a sphere–sphere capacitance, C_{ssa}, can be written in terms of C_{sp} by

$$C_{ssa}(R_1, R_2, s) = \gamma C_{sp}(R_{min}, \gamma s), \tag{15.13}$$

where R_{min} and R_{max} are the smaller and larger of the two radii, and γ is

$$\gamma = \frac{1}{1 + R_{min}/R_{max}} \tag{15.14}$$

15.8.3
Example

To compare the exact and approximate solutions to the sphere–sphere capacitance, we choose two cases. In the first case, shown in Figure 15.7(a), where $R_1 = R_2 = 100$ nm, the exact and approximate solutions are practically identical. In the second case, shown in Figure 15.7(b), where $R_1 = 100$ nm and

$R_2 = 10$ nm, the exact and approximate solutions are quite close as well. The approximate solution reproduces the exact one when one of the radii of the spheres is much larger than the other, and when they are equal. The simplicity of the approximate solution, therefore, merits its use.

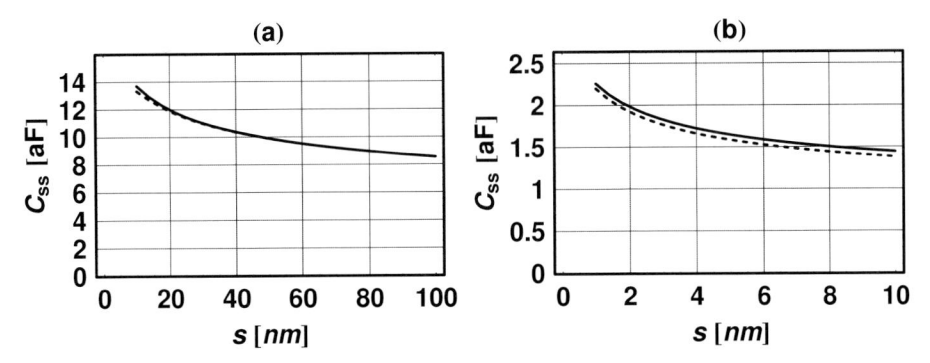

Fig. 15.7 The sphere–sphere capacitance, C_{sp}, as a function of their separation, s, for (a) $R_1 = R_2 = 100$ nm and (b) $R_1 = 100$ nm and $R_2 = 10$ nm.

15.9
Electrostatic Force

The capacitance of a conducting tip of a scanning probe microscope, placed next to a flat conducting sample, can be approximated by that of a conducting sphere placed next to a conducting plane. Applying a bias v between these two bodies gives rise to an electrostatic force, F_{sp}, given by

$$F_{sp} = -\frac{1}{2}\frac{\partial}{\partial s}C_{sp}v^2. \qquad (15.15)$$

Figure 15.8 shows the electrostatic force, F_{sp}, exerted on a sphere with radius $R = 25$ (solid line) and 100 nm (dashed line), separated from a flat, conducting plane by a distance s, with a bias of $v = 1$ V applied between these two. Note that the electrostatic force plays an important role in scanning tunneling microscopy and conducting-tip atomic force microscopy, and may have to be taken in account.

▶ **Exercises for Chapter 15**

1. Use the method of images to show that Eq. (15.6) and Eq. (15.7) yield the proper value for the sphere–plane capacitance.

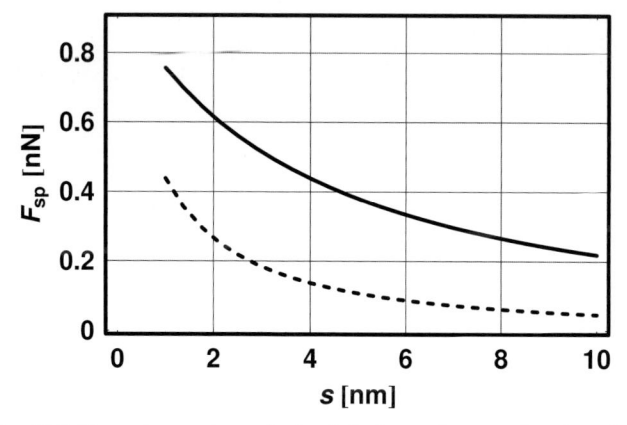

Fig. 15.8 The sphere–plane electrostatic force, F_{sp}, as a function of their separation, s, for $R = 25\,\text{nm}$ (solid line) and $100\,\text{nm}$ (dashed line).

2. Verify that the solution of the two-sphere capacitance reproduces the sphere–plane capacitance if one of the radii is much larger than the other, and if they are equal.
3. Calculate, code, and plot the potential and field for a sphere–sphere geometry.

References

1. P. Lorrain and D. Corson, *Electromagnetic Fields and Waves*, 2nd Edition, W. H. Freeman, San Francisco, 1970.
2. Guanming Lai, private communication.

16
Near-Field Optics

Highlights

1. Bethe–Bouwkamp solution to light diffracted by a circular aperture in a perfectly conducting screen
2. Plots of electric fields and radiation intensities in the near- and far-field
3. Electric fields across a hexagonal array of subwavelength apertures

Abstract

Near-field scanning optical microscopy (NSOM) is concerned with an electromagnetic wave diffracted by a subwavelength aperture with radius a positioned at the end of a metal-coated fiber or hollow pipette. This chapter presents a simplified situation in which a plane, linearly polarized electromagnetic wave is incident normally on a circular aperture in a perfectly conducting screen. The diffraction of this wave, whose wavelength is λ, from the aperture whose radius a is smaller than λ, is presented in two regions. The first, which is based on the radiation of a magnetic dipole, is applicable for aperture–sample distances z such that $z > a$. The second, originally derived by Bethe and Bouwkamp, is applicable for $0 < z < a$. Expressions for the associated electric and magnetic fields and radiation intensities are presented for both regions together with plots of interest. Finally, electric fields across a hexagonal array of subwavelength apertures, fabricated in a perfect conductor, are calculated and plotted.

16.1 Introduction

In a typical near-field scanning optical microscope (NSOM), light is guided inside a metal-coated fiber or a hollow pipette that are terminated by a subwavelength aperture. The aperture, which is in the $\hat{x}$–$\hat{y}$ plane, is placed at a distance $z < a$ away from the surface of a sample across which it is raster scanned. The light diffracted by the aperture interacts with a subwavelength area across a sample. The interaction can involve, for example, absorption of the light diffracted by the aperture by opaque features across the sample, or excitation of fluorescing molecules adsorbed across the sample. In both cases, the far-field radiation emanating from the sample and detected by a photodetector, produces a subwavelength-resolution image of these features.

The unique characteristics of NSOM include (a) a high concentration of optical radiation in a subwavelength region that provides a high signal-to-noise ratio image relative to that obtained by conventional optical microscopy, and (b) the generation of near-field radiation with an altered state of polarization, in contrast to conventional optical microscopy whose state of polarization stays unchanged. Bethe's original solution to the problem of electromagnetic fields diffracted by a small circular aperture, followed by the work of Bouwkamp, assumes a perfectly conducting plane, and therefore neglects skin effects. In their theory, the incident electromagnetic fields generate oscillating currents at the circumference of the aperture, whose radiation fields interfere with the incident field. The theory shows that the resultant radiation in the far field is dipolar in nature. In proximity to the aperture, namely for $z < a$, the diffracted electromagnetic fields from an $\hat{x}$-polarized incident field are found to differ from those obtained using Kirchoff's method. In particular, the diffracted fields have not only the original plane polarized component of the incident field, but also components in the two other directions that decay farther away from the aperture. A manifestation of the polarization characteristics was demonstrated by Betzig, who used an NSOM to excite fluorescing molecules across the surface of a sample. An extension of Bethe–Bouwkamp solution was given by Grober using Fourier optics, a method that enables one to determine the diffracted fields away from the aperture, from a known solution at the aperture plane.

This chapter presents the far-field solution to the electromagnetic fields in terms of magnetic dipole radiation and the near-field solution of Bethe and Bouwkamp. Electric and magnetic fields, Poynting vectors, and intensities are given for both cases and plots of interest presented. Finally, electric fields across a hexagonal array of subwavelength apertures, fabricated in a perfect conductor, are calculated and plotted. Analysis of the distribution of these fields sheds light on a variety of interaction such as Raman and fluorescence, for example, that can take place between a sharp, conducting tip of an AFM and molecular species across the surface of a sample. We will be using Gaussian, rather than MKSA units for ease of making contact with the work cited in the literature. A typical choice of experimental parameters, given in Table 16.1, will be used in all the example presented in this chapter.

16 Near-Field Optics

Table 16.1 A typical choice of experimental parameters used in all the examples presented in this chapter.

$\lambda = 500\,\text{nm}$	
$a = 50\,\text{nm}$	

16.2
Far-Field Solution

16.2.1
Vector Potential

As noted in the Bethe–Bouwkamp theory, the far-field radiation of light diffracted by a circular aperture with radius a can be modeled as that generated by an oscillating magnetic dipole whose moment, m_0, is

$$m_0 = -\frac{2}{3\pi}a^3 H_0. \tag{16.1}$$

Here, H_0 is the magnetic field of the incident light, which we assume to be equal to unity. The dipole is positioned in the aperture's $\hat{x}$–$\hat{y}$ plane, pointing along the $\hat{y}$ direction, which is perpendicular to the direction of polarization of the light incident on the aperture. The vector potential of the magnetic dipole, A_d, is given by

$$A_d = \left\{0, 0, \frac{e^{i(kr-\omega t)}}{r^2} m_0(1 - ikr)\sin\theta\right\}. \tag{16.2}$$

using the spherical coordinate system $\{\hat{r}, \hat{\theta}, \hat{\phi}\}$. It will be convenient to rewrite A_d in terms of an amplitude, $A_{d,0}$

$$A_{d,0} = m_0\sqrt{1 + k^2 r^2}\sin\theta, \tag{16.3}$$

and phase

$$\psi = kr - \omega t + \arctan(-kr). \tag{16.4}$$

16.2.2
Electric Field

The electric field generated by the oscillating magnetic dipole, E_d, is obtained from A_d using

$$E_d = -\frac{1}{c}\frac{d}{dt}A_d, \tag{16.5}$$

with an amplitude, $E_{d,0}$, given by

$$E_{d,0} = \left\{ 0, 0, \frac{\omega m_0 \sqrt{1+k^2 r^2}}{cr^2} \sin\theta \right\}. \tag{16.6}$$

The electric field, in Cartesian coordinates, is clearly in the $\hat{x}$–$\hat{z}$ plane. Transforming the electric field from spherical to Cartesian coordinate systems yields its three Cartesian components

$$E_{d,x} = \sin\phi E_d, \tag{16.7}$$
$$E_{d,z} = \cos\phi E_d, \tag{16.8}$$

and

$$E_{d,y} = 0. \tag{16.9}$$

16.2.3
Electric Vector Field in the $\hat{x}$–$\hat{z}$ Plane

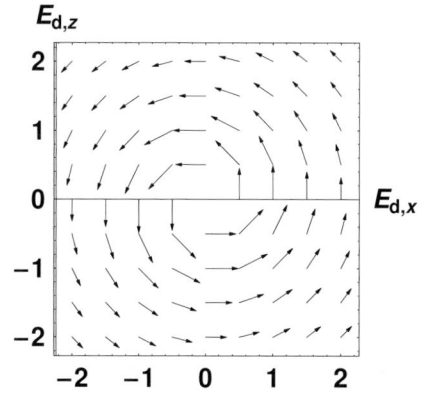

Figure 16.1 shows the electric field, E_d, radiated by the magnetic dipole in the $\hat{x}$–$\hat{z}$ plane at a distance $y = a$. The electric field vectors, as expected, are circulating around the direction of the magnetic dipole which is in the $\hat{y}$ direction.

Fig. 16.1 The electric field, E_d, radiated by the magnetic dipole in the $\hat{x}$–$\hat{z}$ plane at a distance $y = a$.

16.2.4
Electric Vector Field in the $\hat{x}$–$\hat{y}$ Plane

Figure 16.2 shows the electric field radiated by the magnetic dipole in a plane parallel to that of the aperture and at a distance $z = a$. Here the electric field lines are along the $\hat{x}$ direction, as expected.

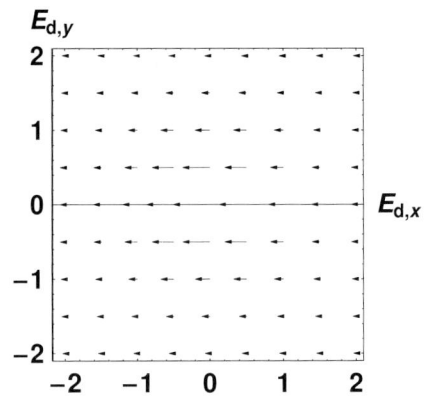

Fig. 16.2 The electric field, E_d, radiated by the magnetic dipole in a plane parallel to that of the aperture at a distance $z = a$.

Using a density plot of field components is (literally) illuminating, for it gives a feel for the light diffracted by the aperture. Figure 16.3 is a density plot of $|E_{d,x}|$ in the aperture's $\hat{x}$–$\hat{y}$ plane at a distance $z = a$ showing a round pattern.

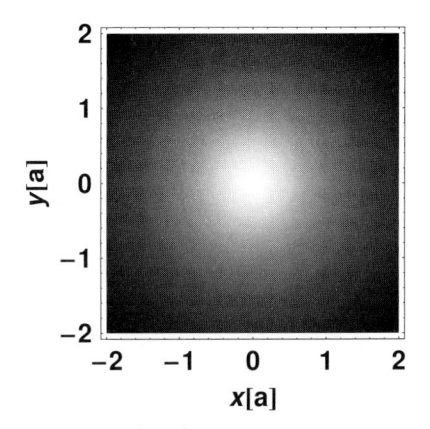

Fig. 16.3 A density plot of $|E_{d,x}|$ in the aperture's $\hat{x}$–$\hat{y}$ plane at a distance $z = a$.

Figure 16.4 is a density plot of $|E_{d,z}|$ in the aperture's $\hat{x}$–$\hat{y}$ plane at a distance $z = a$, showing two distinct lobes.

16.2.5
Magnetic Field

The magnetic field, H_d, is obtained from the curl of the vector potential, A_d, by

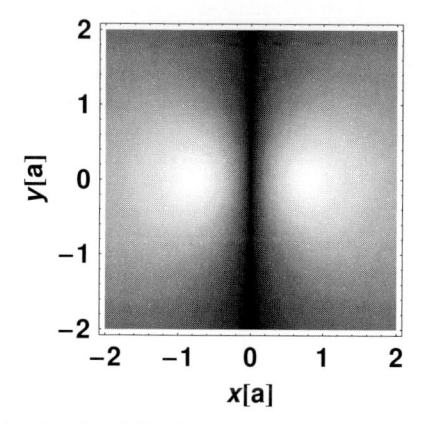

Fig. 16.4 A density plot of $|E_{d,x}|$ in the aperture's $\hat{x}$–$\hat{y}$ plane at a distance $z = a$.

$$H_d = \text{curl } A_d. \tag{16.10}$$

The three Cartesian components of H_d are

$$H_{d,x} = \frac{2m_0\sqrt{1 + k^2 r^2}\cos\theta}{r^3}, \tag{16.11}$$

$$H_{d,y} = \frac{m_0(1 - ik^3 r^3)\sin\theta}{r^3\sqrt{1 + k^2 r^2}}, \tag{16.12}$$

and

$$H_{d,z} = 0. \tag{16.13}$$

16.2.6
Poynting Vector and Intensity

The Poynting vector, S_d, obtained from the cross product of the electric and magnetic fields, E_d and H_d, respectively, is

$$S_d = \frac{c}{4\pi}[E_d \times H_d]. \tag{16.14}$$

The intensity of the radiation, I_d, is obtained by the scalar product of the Poynting vector with itself and taking the square root of the product,

$$I_d = \sqrt{S_d \cdot S_d}. \tag{16.15}$$

Recall that the aperture is in the $\hat{x}$–$\hat{y}$ plane and that the electric field of the normally incident light is polarized in the $\hat{x}$ direction, resulting in a magnetic

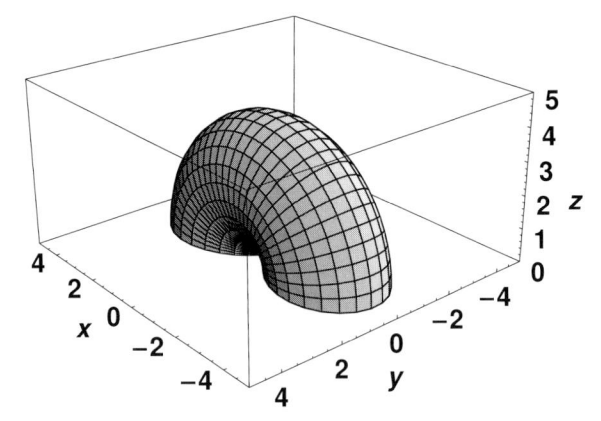

Fig. 16.5 The intensity of the radiation pattern of the magnetic dipole, I_d, at a radius r away from the aperture.

moment pointing in the $\hat{y}$ direction. Figure 16.5 shows the intensity of the radiation pattern of the magnetic dipole, I_d, at a radius r away from the aperture, which vanishes along the $\hat{y}$ axis.

16.2.7
Intensity in the $\hat{x}$–$\hat{z}$ Plane

The intensity of the radiation of the magnetic dipole, I_d, projected onto the $\hat{x}$–$\hat{z}$ plane, at a distance $y = a$, is shown in Figure 16.6, clearly indicating that it vanishes along the $\hat{y}$ axis.

Fig. 16.6 A density plot of the intensity of the radiation of the magnetic dipole, I_d, projected onto the $\hat{x}$–$\hat{z}$ plane at a distance $y = a$.

16.2.8
Intensily in the $\hat{x}$–$\hat{y}$ Plane

The intensity of the radiation of the magnetic dipole, I_d, projected onto the $\hat{x}$–$\hat{y}$ plane, at a distance $z = a$, is shown in Figure 16.7.

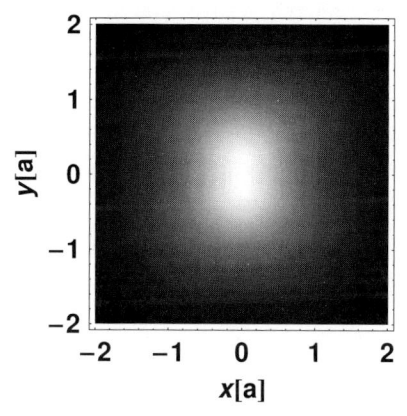

Fig. 16.7 A density plot of the intensity of the radiation of the magnetic dipole, I_d, projected onto the $\hat{x}$–$\hat{y}$ plane, at a distance $y = a$.

Figure 16.1 to Fig. 16.7 showed the electric field components and intensity of the radiation of the magnetic dipole that model the far-field region of light diffracted by the circular aperture. We will now treat the radiation emanating from the aperture in the near-field region, using the Bethe–Bouwkamp solution.

16.3
Near-Field Solution

16.3.1
Transformation

The electric and magnetic fields in the near-field region, namely, in the range $0 < z < \lambda/10$, were derived by Bethe and Bouwkamp. Bouwkamp, in his derivation, used an oblate–spheroidal coordinate system defined by three variables, a, u, and v. These are related to x, y, and z by

$$x = a\sqrt{(1-u^2)(1+v^2)} \cos\phi, \tag{16.16}$$

$$y = a\sqrt{(1-u^2)(1+v^2)} \sin\phi, \tag{16.17}$$

and

$$z = auv. \tag{16.18}$$

It will be handy to present Bouwkamp's results in a Cartesian coordinate system, which is readily accomplished by solving for a function f such that $\{a, u, v\} = f\{x, y, z\}$.

16.3.2
Electric Field

The three Cartesian components of the electric field diffracted by the aperture, E_d, were given by Bethe and Bouwkamp as

$$E_{a,x} = kz - \frac{2}{\pi}kau\left(1 + v\arctan v + \frac{1}{3(u^2+v^2)} + \frac{x^2-y^2}{3a^2(u^2+v^2)(1+v^2)^2}\right), \tag{16.19}$$

$$E_{a,y} = -\frac{4}{3\pi}\frac{kxyu}{a(u^2+v^2)(1+v^2)^2}, \tag{16.20}$$

and

$$E_{a,z} = \frac{4}{3\pi}\frac{kxv}{(u^2+v^2)(1+v^2)}. \tag{16.21}$$

The parameters u and v can be eliminated through the use of the function f for plotting purposes.

16.3.3
Electric Field in the $\hat{x}$–$\hat{y}$ Plane

Figure 16.8 shows $|E_{a,x}|$, which is parallel to the aperture's plane, for three distances, (a) $z = a/100$, (b) $z = a/10$, and (c) $z = a$.

Note that the shape of the field shown in Figure 16.8(c) resembles that which is produced by a dipole radiation in the far field. The square of the fields at the center of Figure 16.8(a–c), given in Table 16.2, show that they do not have an obvious dependence on the distance z.

Table 16.2 The square of the fields at the center of Figure 16.8(a–c).

z	$|E_{a,x}|^2$
$a/100$	0.277 81
$a/10$	0.223 888
a	0.023 258 5

Figure 16.9 shows $|E_{a,y}|$, which is parallel to the aperture's plane, for the same three distances used in Figure 16.8.

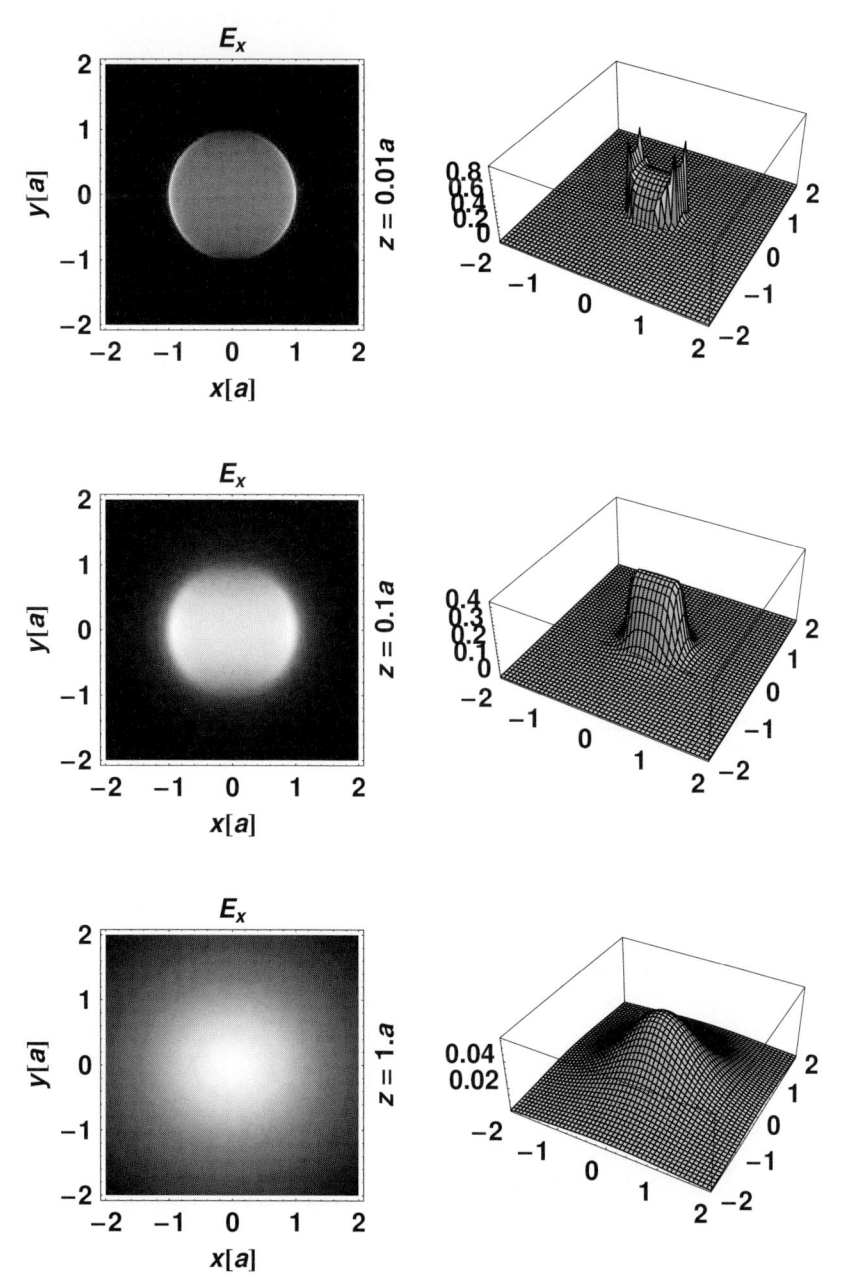

Fig. 16.8 Density plots of $|E_{a,x}|$ for three distances, $z = a/100$ (top), $z = a/10$ (center), and $z = a$ (bottom).

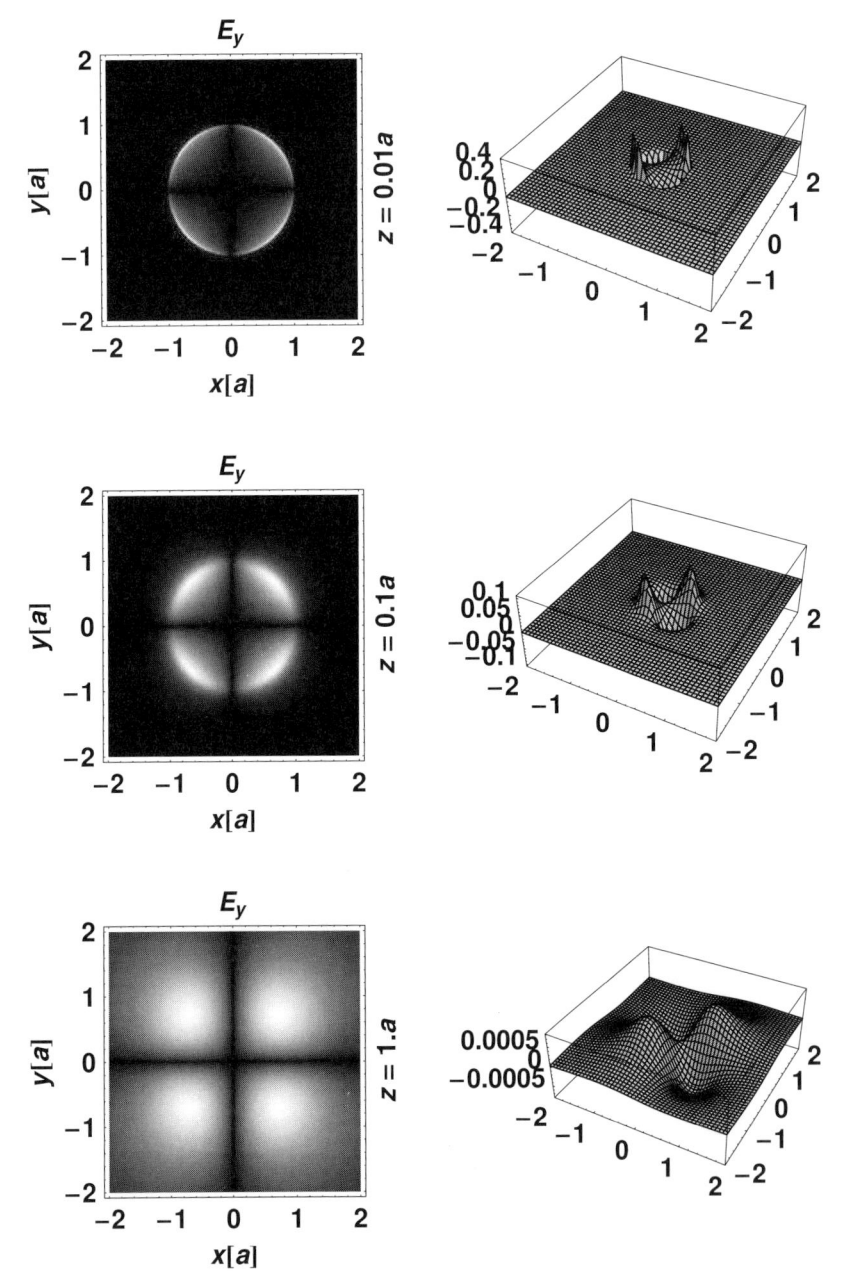

Fig. 16.9 Density plots of $|E_{a,y}|$, for the same three distances used in Figure 16.8.

Here, we cannot compare the fields to those produced by the dipole whose radiation does not have a component in this direction. Figure 16.10 shows $|E_{a,z}|$, which is parallel to the aperture's plane, for the same three distances used in Figure 16.8.

Figure 16.8 to Figure 16.10 are useful when probing the excitation of molecules adsorbed on the surface of a sample by a near-field microscope. The unique shape of these fields will, therefore, play a major role in the efficiency of excitation of such molecules, as shown by Betzig.

16.3.4
Magnetic Field

The three cartesian component of the magnetic field diffracted by the aperture, H_d, were given by Bethe and Bouwkamp as

$$H_{a,x} = -\frac{4}{\pi} \frac{xyv}{a^2(u^2+v^2)(1+v^2)^2}, \tag{16.22}$$

$$H_{a,y} = 1 - \frac{2}{\pi}\left(\arctan v + \frac{v}{u^2+v^2}\right) + \frac{2}{\pi} \frac{(x^2-y^2)v}{a^2(u^2+v^2)(1+v^2)^2}, \tag{16.23}$$

and

$$H_{a,z} = -\frac{4}{\pi} \frac{ayu}{a^2(u^2+v^2)(1+v^2)}. \tag{16.24}$$

The parameters u and v can be eliminated through the use of the function f for plotting purposes.

16.3.5
Poynting Vector and Intensity

The Poynting vector, S_a, is defined in terms of E_a and H_a by the vector product

$$S_a = \frac{c}{4\pi}[E_a \times H_a]. \tag{16.25}$$

The radiation intensity, I_a, is the absolute value of the Poynting vector,

$$I_a = |S_a|. \tag{16.26}$$

Figure 16.11 shows I_a in a plane parallel to the aperture for three distances, (a) $z = a/100$, (b) $z = a/10$, and (c) $z = a$.

The maximum value of I_a in Figure 16.11(a–c) are given in Table 16.3, and as before, there is no simple relationship between the intensity and the distance.

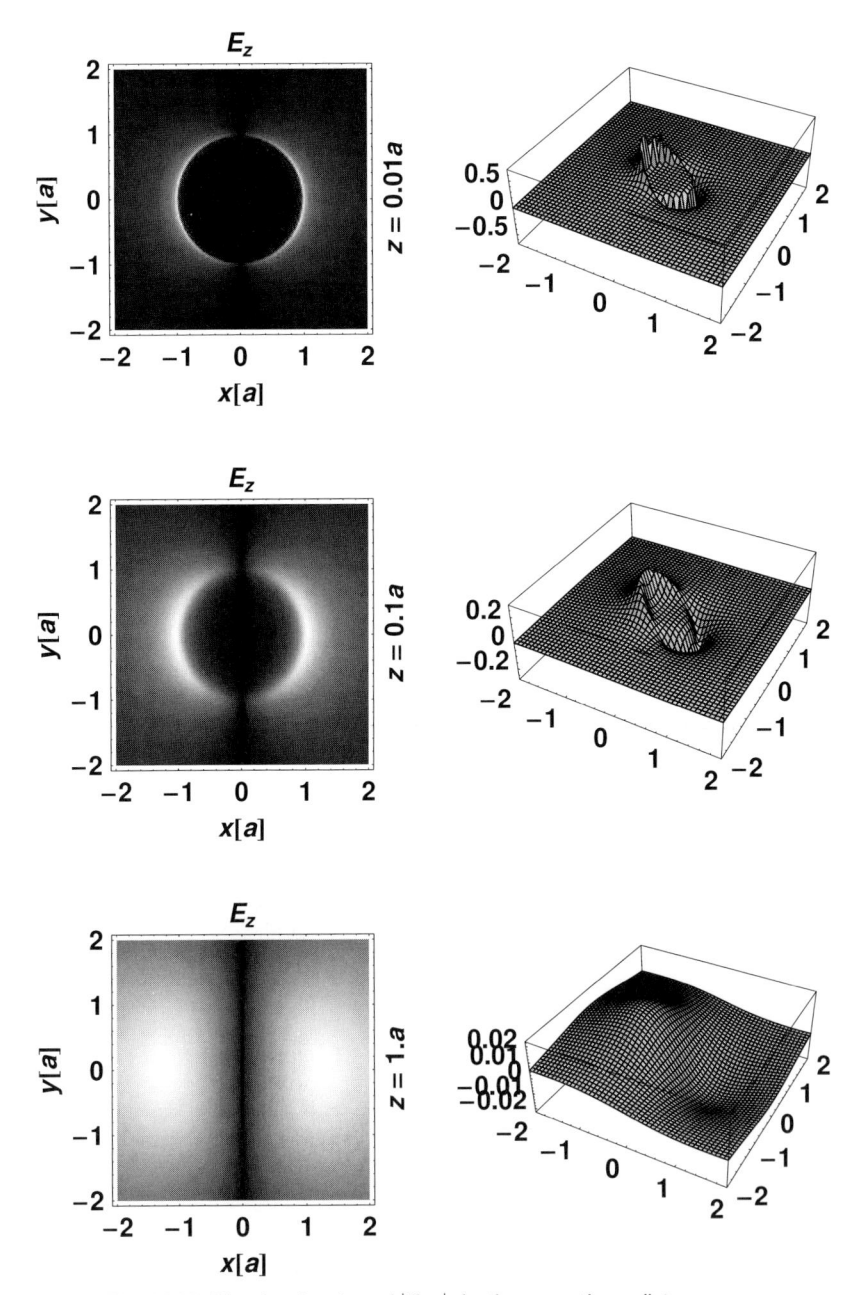

Fig. 16.10 The density plots of $|E_{a,z}|$, for the same three distances used in Figure 16.8.

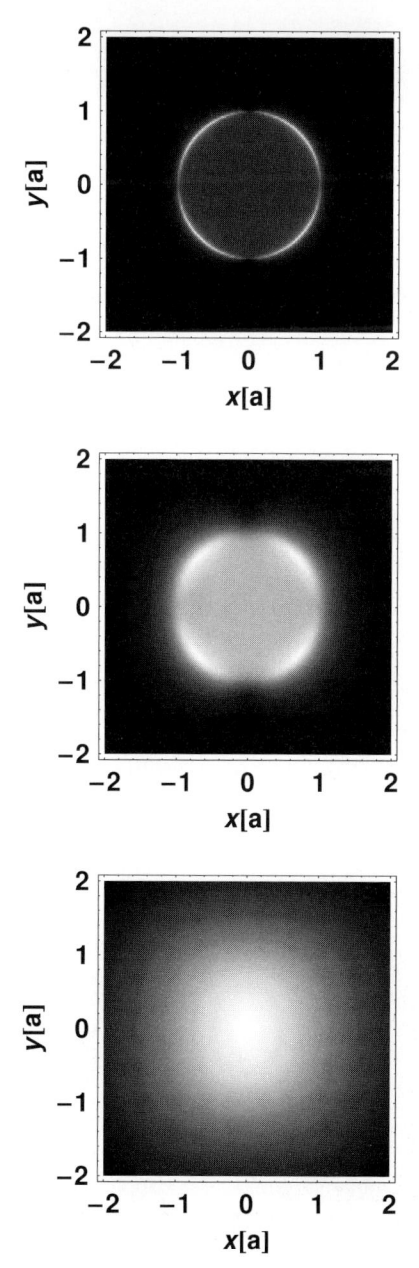

Fig. 16.11 Density plots of the radiation intensity, I_a, parallel to the aperture's plane for three distances, (a) $z = a/100$, (b) $z = a/10$, and (c) $z = a$.

Table 16.3 The maximum value of I_a in Figure 16.11(a–c) for three distances, (a) $z = a/100$, (b) $z = a/10$, and (c) $z = a$.

z	I_a
$a/100$	12.4142
$a/10$	9.860 47
a	0.661 048

16.4
Discussion of the Models

16.4.1
Electric Field

The main deficiency in the solution of the diffracted electromagnetic fields presented in this chapter is that we did not take into account the skin depth of the conducting screen surrounding the aperture. The predicted intensity of the diffracted light will therefore not be accurate for apertures whose radius is of the order of the skin depth, namely about 12.5 nm. Nevertheless, the lack of a better analytic solution that does take into account the skin depth has made the solution offered by Bethe and Bouwkamp the tool of choice for analyzing near-field optical microscopy. It is of interest to compare the near- and far-field solutions at a distance of $z = a = \lambda/10$, for example. Figure 16.12 and Figure 16.13 show (a) $|E_{d,x}|$ (solid line) and $|E_{a,x}|$ (dashed line), and (b) $|E_{d,z}|$ (solid line) and $|E_{a,z}|$ (dashed line), respectively, at $y = 0$ along the $\hat{x}$ direction. We find that while the magnitude of the fields at $x = 0$ are close in their value, their shapes and widths differ somewhat.

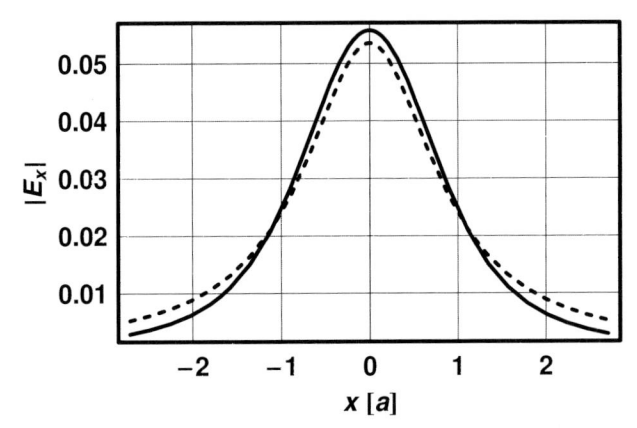

Fig. 16.12 $|E_{d,x}|$ (solid line) and $|E_{a,x}|$ (dashed line) at $y = 0$ and $z = \lambda/10$ along the $\hat{x}$ direction.

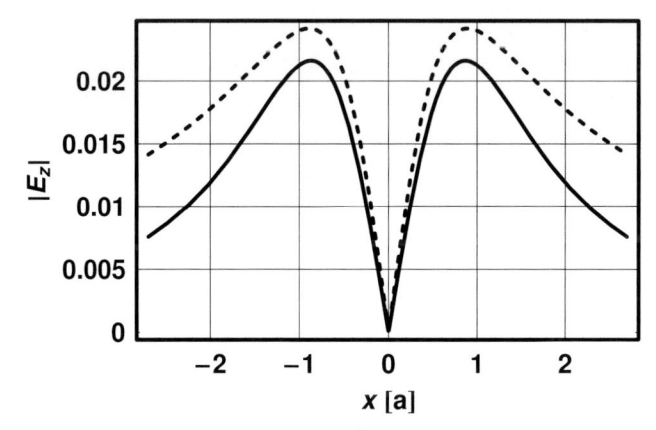

Fig. 16.13 $|E_{d,z}|$ (solid line) and $|E_{a,z}|$ (dashed line) at $y = 0$ and $z = \lambda/10$ along the $\hat{x}$ direction.

16.4.2
Intensity

Figure 16.14 shows the radiated intensity from a magnetic dipole, I_d, (solid line) and an aperture, I_a, (dashed line) at $y = 0$ along the $\hat{x}$ direction.

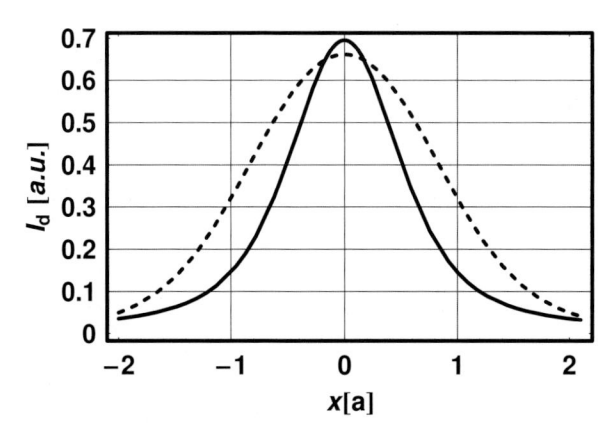

Fig. 16.14 The radiated intensity from a magnetic dipole, I_d, (solid line) and an aperture, I_a, (dashed line) at $y = 0$ along the $\hat{x}$ direction.

As in Figures 16.12 and 16.13, the results at $x = 0$ are similar in both models, while the near-field solution yields a wider curve. It is instructive to note that the total intensity diffracted by the aperture, I_a, is

$$I_a = \frac{c}{27\pi^2} k^4 a^6 (4H_a^2 + E_a^2), \qquad (16.27)$$

while the ratio of the transmitted to the incident intensities per unit area, T, is

$$T = \frac{64\pi^2}{27}\left(\frac{a}{\lambda}\right)^4. \tag{16.28}$$

Note that $I_a \propto a^6$ while $T \propto a^4$, because the latter assumes an incident intensity that has the same area as that of the aperture. We already mentioned that at $z = \lambda/10$, the Bethe–Boukamp solution differs from that of a magnetic dipole. Clearly, the near-field solution is applicable only for $z < a$, while the far-field solution is applicable only for $z > \lambda$. A general solution to the diffracted fields in terms of Fourier optics, developed by Grober, covers the whole range of aperture–sample distances, and in particular it bridges the gap between the near- and far-field solutions in the range $a < z < \lambda$. The coding of his general solution, which is somewhat similar to the one presented in this chapter, is left as an exercise.

16.5
Scattered Electric Fields Around Patterned Apertures

Figure 16.15 is a schematic of two hexagonal arrays of subwavelength apertures, fabricated in a perfect conductor, rotated at an angle of $\phi = \pi/12$ relative to each other and to an incident beam of light polarized in the $\hat{x}$ direction.

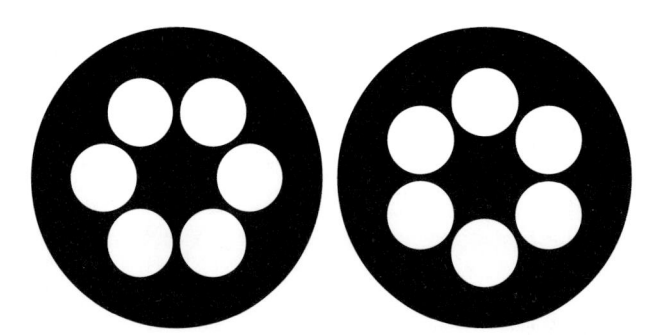

Fig. 16.15 A schematic of a two hexagonal arrays of subwavelength apertures, fabricated in a perfect conductor, rotated at an angle of $\phi = \pi/12$ relative to each other and to an incident beam of light polarized in the $\hat{x}$ direction.

Figure 16.16 and Figure 16.17 show the scattered electric fields E_x, E_y, and E_z, and the total field, $E_T = E_x + E_y + E_z$, for an $\hat{x}$-polarized beam of light that is incident normally on the hexagonal apertures shown in Figure 16.15 on the left- and right-hand side, respectively. The fields are shown at a distance $z = a/100$, $a/10$, and a. The distribution of the fields differs because of the different symmetries of the hexagonal structures relative to the polarization of the light. In each case, however, the total field in these two cases is the same for $z > a$.

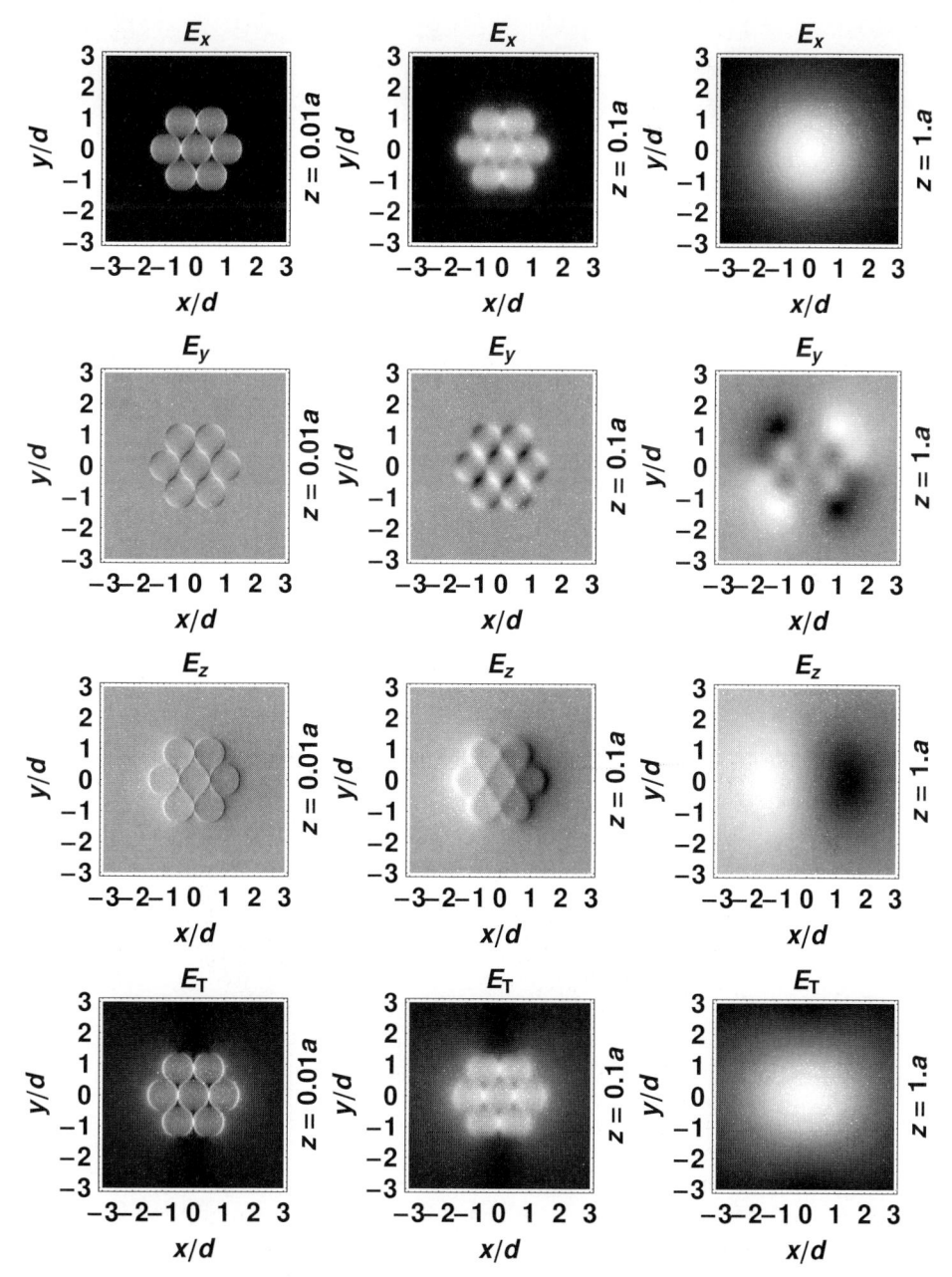

16.5 Scattered Electric Fields Around Patterned Apertures | 239

Fig. 16.16 The scattered electric fields E_x, E_y, and E_z, and the total field, $E_T = E_x + E_y + E_z$, for an $\hat{x}$-polarized beam of light that is incident normally on the hexagonal apertures shown in Figure 16.15 on the left-hand side. The fields are shown at a distance $z = a/100$, $a/10$, and a.

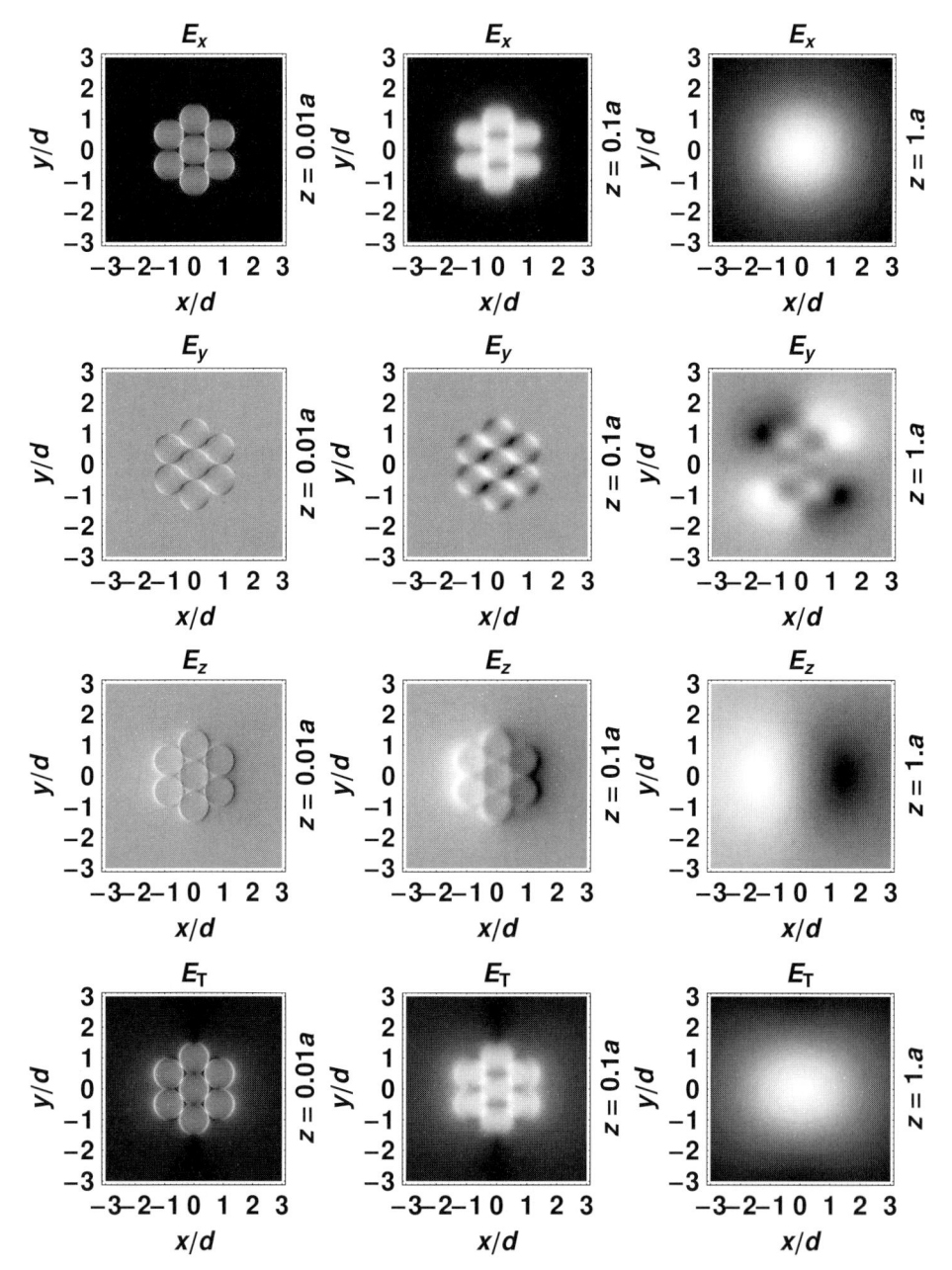

Fig. 16.17 The scattered electric fields E_x, E_y, and E_z, and the total field, $E_T = E_x + E_y + E_z$, for an $\hat{x}$-polarized beam of light that is incident normally on the hexagonal apertures shown in Figure 16.15 on the right-hand side. The fields are shown at a distance $z = a/100$, $a/10$, and a.

▶ **Exercises for Chapter 16**

1. Transform the fields, Poynting vectors, and intensities to MKSA units.
2. Calculate the fields, Poynting vectors, and intensities for a beam of light having circular, rather than linear polarization.
3. Extend the Bethe–Bouwkamp model to the range $a < z < \lambda$, using Fourier optics [5].

References

1 H. A. Bethe, "Theory of diffraction by small holes," Phys. Rev. **66**, 163 (1944).
2 J. D. Jackson, *Classical Electrodynamics*, Wiley, New York, 1975.
3 C. J. Bouwkamp, "On Bethe's theory of diffraction by small holes," Philips Res. Rep. **5**, 321 (1950).
4 E. Betzig and R. J. Chichester, "Single molecules observed by near-field optical microscopy," Science **262**, 1422 (1993).
5 R. D. Grober, T. Rutherford, and T. D. Harris, "Modal approximation for the electromagnetic field of a near-field optical probe," Appl. Opt. **35**, 3488 (1996).

17
Constriction and Boundary Resistance

Highlights

1. Electrical properties of metals modeled as a free electron gas
2. Electrical and thermal constriction resistance
3. Electrical and thermal boundary resistance

Abstract

Many metals behave as a free electron gas and their electrical and thermal characteristics can be described by simple equations that reproduce experimental results rather accurately. Results for these characteristics are presented in figures where the values for Cu, Al, and Pt are depicted. A model for electrical transport through a constriction for the whole range of Knudsen numbers is presented together with examples. The effect of a diffuse-mismatch-like boundary on the transport of electrons is given, and the results are found to be identical to those of transport through a high Knudsen number constriction. Several important questions regarding the results presented in this chapter are raised and left to the reader for further exploration.

17.1
Introduction

This chapter describes the electrical and thermal properties of metals that are modeled like a free electron gas. The second section starts with the connection between electrical and thermal conductivities, κ and σ, respectively, made through the Wiedemann–Franz law using the Lorenz number, N_{Lorenz}. Next, the Fermi velocity, v_F, temperature, T_F, and k-vector, k_F, are presented, followed by the definitions of the number of electrons per volume, N_e, and the electronic density of states per volume, $\mathcal{D}_e$. Finally, the mean free path, ℓ, the ratio of ℓ/σ, and the electronic heat capacitance per volume, C_e, are described. Each of these concepts is accompanied by a figure that describes its dependence on the Fermi energy, $\mathcal{E}_F$, given in units of eV. The three metals, Cu, Al, and Pt are used extensively throughout this chapter as typical examples that cover the range of Fermi energies of most other metals.

The third section deals with the case where two semi-infinite metallic media are separated by a barrier having a round conducting aperture with radius r. When a voltage is applied between these two media, the resulting current is constricted to flow through the aperture. The resistance due to the presence of the constriction is a strong function of the Knudsen number, $K = \ell/r$. For small values of K, the current is in the Maxwell regime, which is proportional to the radius of the constriction. For large values of K, the current is in the Sharvin regime, which is proportional to the area of the constriction. An approximate solution makes it possible to obtain an analytical expression for the electrical resistance for intermediate values of K that can be applied also to thermal resistance.

The fourth section deals with two semi-infinite media whose thermal energies are mediated by phonons. These media are separated by a continuous boundary that acts as a totally diffuse scatterer of phonons. Phonons belonging to each medium, when incident on this boundary, are scattered isotropically in all directions, including the adjacent medium. The transmission through such a boundary is described by the diffuse mismatch model (DMM), which is contrasted from the elastic mismatch model (EMM), where phonons obey acoustic Fresnel equations. While the EMM model applies to very low temperatures, the DMM model is more applicable to higher temperatures. When a temperature difference is applied between these two media, the thermal heat flux from one medium to the other encounters a resistance on passage through the boundary. The diffuse mismatch model for phonons can also be applied to two semi-infinite metallic media whose thermal energy is mediated by electrons. The expressions for the thermal and electrical resistances due to the boundary are presented in terms of the Fermi energy of each medium. Both these resistances are proportional to the area of the barrier that separates the two semi-infinite media. The results of this section are accompanied by figures applicable to Cu, Al, and Pt.

It is interesting that the expressions for the thermal and electrical Sharvin resistances are identical to those of the boundary resistance described by the diffuse mismatch model. In other words, for a given area of a constriction separating two semi-infinite metallic media, in the large Knudsen limit, the thermal and electrical resistances act exactly as that of a diffuse scattering boundary. A discussion of this result is presented in the fifth section of this chapter. The end of the chapter contains a set of exercises and a list of references, respectively.

The results obtained in this chapter can be used, for example, when analyzing the injection of electrical or thermal currents from the tip of an AFM into a metallic or dielectric sample. In particular, the voltage and temperature drop across a tip–sample constricted boundary yield values that have to be taken into account when designing a variety of systems, such as a thermal conductivity microscope and probe data storage.

17.2
A Metal as a Free Electron Gas

17.2.1
Lorenz Number, N_{Lorenz}, and the Wiedemann–Franz Law

The Lorenz number, N_{Lorenz}, defined by

$$N_{\text{Lorenz}} = \frac{\pi^2}{3} \frac{K_B^2}{e^2}, \tag{17.1}$$

gives the ratio between the thermal and electrical conductivities of metals,

$$\kappa = N_{\text{Lorenz}} \sigma T. \tag{17.2}$$

Here, K_B is Boltzmann's constant, e the electron charge, and T the temperature. This ratio, which is called the Weidemann–Franz law, is quite constant for many metals. As an example, Table 17.1 gives the Wiedemann–Franz ratio, $\kappa/(\sigma T)$, for Cu, Al, and Pt at room temperature.

Table 17.1 The Wiedemann–Franz ratio, $\kappa/(\sigma T)$, together with the experimentally observed values for Cu, Al, and Pt.

	N_{Lorenz}	Cu	Al	Pt
$\frac{\kappa}{\sigma T}$ $(10^{-8}\,\text{W}\Omega/\text{K}^2)$	2.443 03	2.23	2.44	2.51

17.2.2
Fermi Velocity, v_F

The Fermi velocity, v_F, given by

$$v_F = \sqrt{\frac{2e\mathcal{E}_F}{m_e}}, \tag{17.3}$$

is shown in Figure 17.1 as a function of the Fermi energy, $e\mathcal{E}_F$. Note that appropriate values for Cu, Al, and Pt are depicted in this figure.

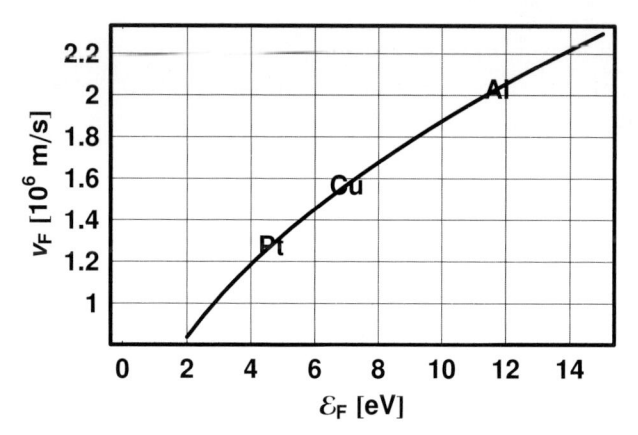

Fig. 17.1 The Fermi velocity, v_F, as a function of the Fermi energy, $e\mathcal{E}_F$.

17.2.3
Fermi Temperature, T_F

The Fermi temperature, T_F, given by

$$T_F = \frac{e\mathcal{E}_F}{K_B},\qquad(17.4)$$

is shown in Figure 17.2 as a function of the Fermi energy, $\mathcal{E}_F$. Note that T_F does not represent an actual temperature; rather it is used as a parameter in calculations.

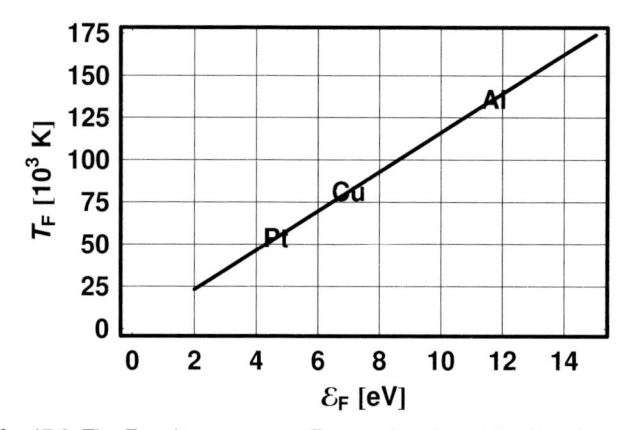

Fig. 17.2 The Fermi temperature, T_F, as a function of the Fermi energy, $\mathcal{E}_F$.

17.2.4
Fermi k-Vector, k_F

The Fermi k-vector, k_F, given by

$$k_F = \frac{2\pi}{h}\sqrt{2m_e e \mathcal{E}_F}, \qquad (17.5)$$

is shown in Figure 17.3 as a function of the Fermi energy, $\mathcal{E}_F$.

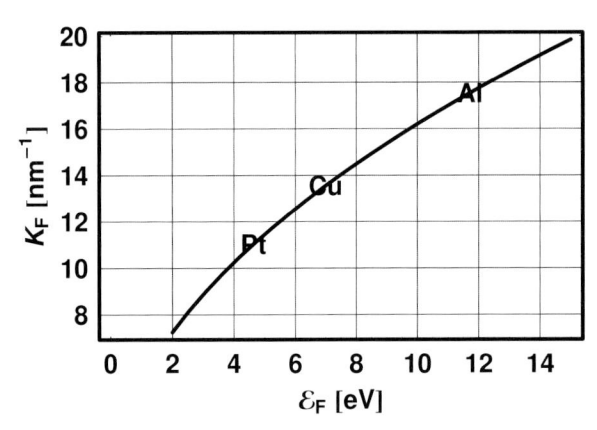

Fig. 17.3 The Fermi k-vector, k_F, as a function of the Fermi energy, $\mathcal{E}_F$.

17.2.5
Electron Density, N_e

The number of electrons per unit volume, N_e, given by

$$N_e = \frac{8\sqrt{2}\pi m_e^{3/2}}{h^3}\frac{2}{3}(e\mathcal{E}_F)^{3/2}, \qquad (17.6)$$

is shown in Figure 17.4 as a function of the Fermi energy, $\mathcal{E}_F$.

17.2.6
Mean Free Path, ℓ

The mean free path of electrons, ℓ, is given in terms of v_F and N_e by

$$\ell = \sigma \frac{m_e v_F}{N_e e^2}. \qquad (17.7)$$

In terms of $\mathcal{E}_F$ and σ, it is given by

$$\ell = \frac{3}{16\pi}\frac{h^3}{e^2 m_e}\frac{1}{e\mathcal{E}_F}\sigma. \qquad (17.8)$$

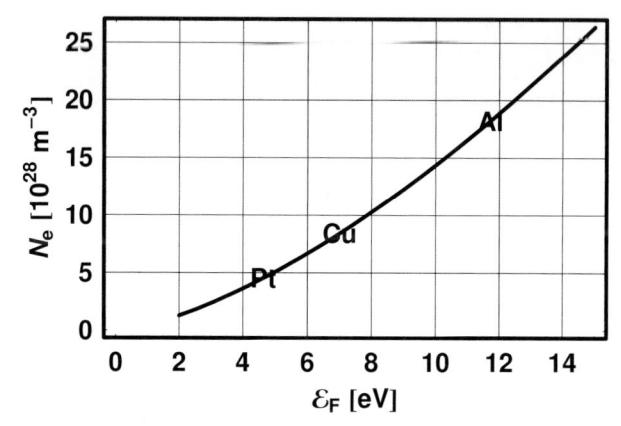

Fig. 17.4 The number of electrons per unit volume, N_e, as a function of the Fermi energy, $\mathcal{E}_F$.

Note that in the presence of defects and boundaries, the effective mean free path, ℓ_{eff}, follows the Matthiessen rule,

$$\frac{1}{\ell_{eff}} = \frac{1}{\ell} + \frac{1}{\ell_{defect}} + \frac{1}{\ell_{boundary}}. \tag{17.9}$$

17.2.7 Ratio of ℓ/σ

The ratio ℓ/σ, in terms of v_F and N_e, is given by

$$\frac{\ell}{\sigma} = \frac{m_e v_F}{N_e e^2}. \tag{17.10}$$

In terms of $\mathcal{E}_F$, it is given by

$$\frac{\ell}{\sigma} = \frac{3}{16\pi} \frac{h^3}{e^2 m_e} \frac{1}{e \mathcal{E}_F}, \tag{17.11}$$

and is shown in Figure 17.5.

17.2.8 Electronic Density of States, $\mathcal{D}_e$

The electronic density of states per volume, $\mathcal{D}_e$, in terms of $\mathcal{E}_F$, is given by

$$\mathcal{D}_e = \frac{1}{2\pi^2}\left(\frac{8\pi^2 m_e}{h^2}\right)^{3/2}(e\mathcal{E}_F)^{1/2}. \tag{17.12}$$

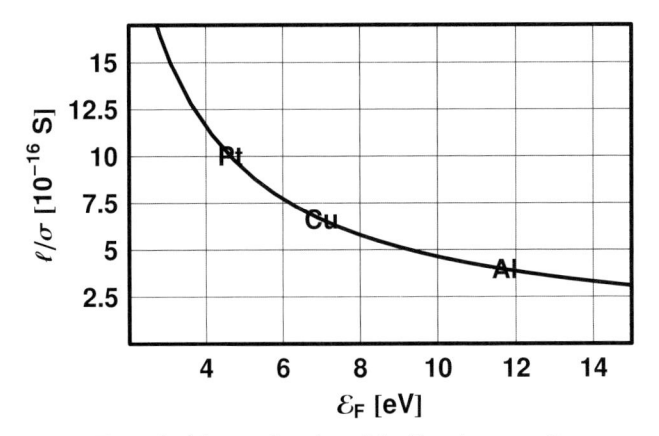

Fig. 17.5 The ratio ℓ/σ as a function of the Fermi energy, $\mathcal{E}_F$.

17.2.9
Electronic Specific Heat, C_e

The electronic specific heat, C_e, is given in terms of $\mathcal{D}_e$ by

$$C_e = \frac{\pi^2}{3} \mathcal{D}_e K_B^2 T, \qquad (17.13)$$

and in terms of N_e and T_F is given by

$$C_e = \frac{1}{2} \pi^2 N_e K_B \frac{T}{T_F}. \qquad (17.14)$$

In terms of $\mathcal{E}_F$, it is given by

$$C_e = \frac{8\sqrt{2}\pi^3}{3} \frac{K_B^2 m_e^{3/2}}{h^3} T(e\mathcal{E}_F)^{1/2}, \qquad (17.15)$$

and is shown in Figure 17.6 as a function of $\mathcal{E}_F$.

17.3
Constriction Resistance

17.3.1
Electrical Resistance in the Maxwell Limit

Consider two semi-infinite metallic media separated by a thin nonconducting barrier having a round aperture (point contact) with radius r, shown in Figure 17.7. The barrier, at which reflection of electrons is assumed, will constrict the electrons to flow through the aperture. The electrical resistance due to this constriction, R_M, in the diffusive regime – namely, the regime where the mean

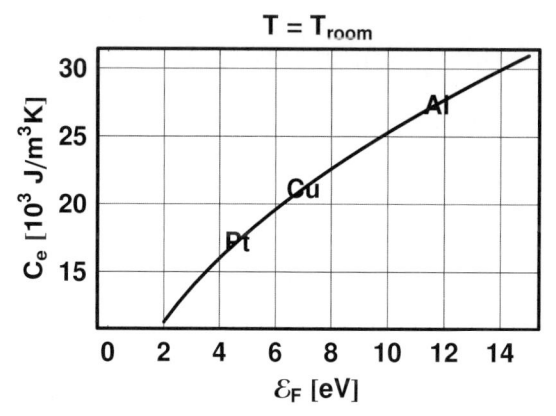

Fig. 17.6 The electronic specific heat, C_e, as a function of the Fermi energy, $\mathcal{E}_F$.

free path of the elastically colliding electrons, ℓ, is much smaller than r – was solved by Maxwell. The driving force acting on the electrons in the vicinity of the constriction varies across the media, generating a distributed electrical resistance. The integral of this distributed resistance (spreading or contact resistance), across which a voltage drop develops, is given by

$$R_M = \frac{1}{\sigma}\frac{1}{2r}, \tag{17.16}$$

for the case in which the two media have the same electrical conductivities, σ. Note that R_M is inversely proportional to the radius, rather than the area of the constriction. Note also that R_M is independent of the thickness of the two bounding media because the major contribution to its value comes from the volume close to the constriction. If the two media have different electrical conductivities, given by σ_1 and σ_2, one uses the replacement

$$\sigma \rightarrow \frac{2}{1/\sigma_1 + 1/\sigma_2}. \tag{17.17}$$

In a metal, R_M is given in terms of ℓ, $\mathcal{E}_F$, and r by

$$R_M = \frac{3}{16\pi}\frac{h^3}{e^2 m_e}\frac{1}{e\mathcal{E}_F \ell}\frac{1}{2r}, \tag{17.18}$$

for identical media. For dissimilar media, one uses the replacement

$$\mathcal{E}_F \ell \rightarrow \frac{2}{\frac{1}{\mathcal{E}_{F_1}\ell_1} + \frac{1}{\mathcal{E}_{F_2}\ell_2}}. \tag{17.19}$$

It should be noted that Eq. (17.16) applies to phonons as well, for which κ replaces σ and ℓ refers to their mean free path. Figure 17.8 shows R_M as a

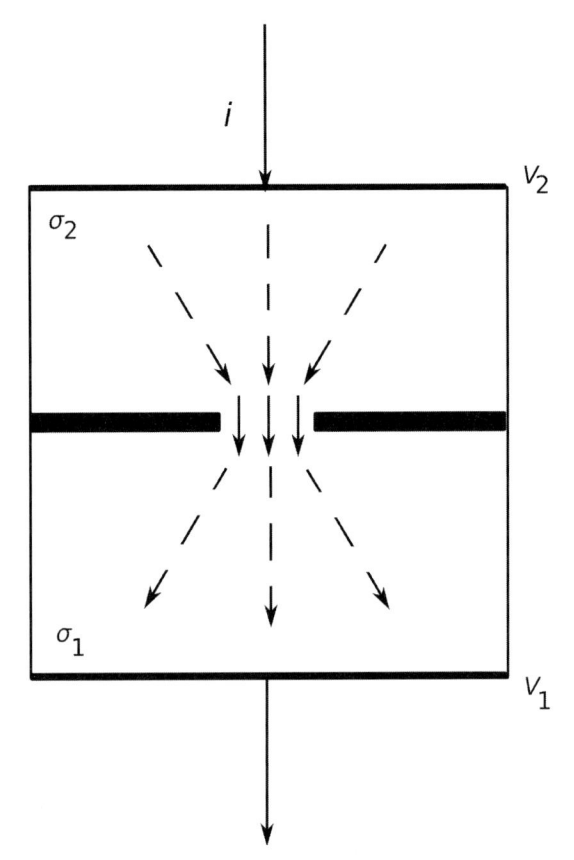

Fig. 17.7 A schematic diagram of two semi-infinite metallic media having electrical conductivities σ_1 and σ_2, separated by a thin nonconducting barrier (denoted by two thick solid lines) having a round aperture with radius r. The two media are held at the voltages V_1 and V_2, resulting in a current flow, i. The direction and length of the arrows indicate the direction of the flow of electrons and the magnitude of their mean free path, respectively. Note that here the mean free path ℓ is much smaller than r.

function of the radius of the constriction for identical bounding media. Here, the solid, dashed, and dotted lines refer to Cu, Al, and Pt media, respectively.

It is important to find out the minimum thickness of each medium for which Eq. (17.16) (for electron or phonons) is valid. To that end, consider a perfectly conducting disk with radius r contacting a metallic cladding film with thickness h and electrical conductivity σ_c, which is deposited on a semi-infinite substrate with electrical conductivity σ_s. The electrical resistance of this structure, R_{cs}, in terms of the parameter K_1 given by

17.3 Constriction Resistance

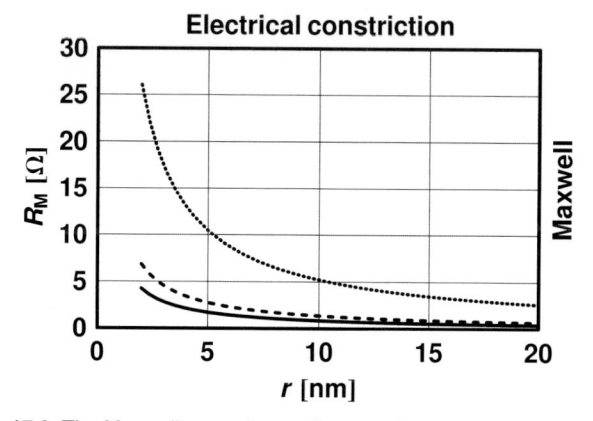

Fig. 17.8 The Maxwell spreading resistance, R_M, as a function of the radius of the constriction, r. Here, the solid, dashed, and dotted lines refer to Cu, Al, and Pt media, respectively.

$$K_1 = \frac{1 - \frac{\sigma_s}{\sigma_c}}{1 + \frac{\sigma_s}{\sigma_c}}, \tag{17.20}$$

is given by

$$R_{cs} = \frac{1}{\pi \sigma_c r} \int_0^\infty \frac{1 + K_1 e^{-2\zeta h_c / r}}{1 - K_1 e^{-2\zeta h_c / r}} J_1(\zeta) \frac{\sin \zeta}{\zeta^2} d\zeta, \tag{17.21}$$

where $J_1(\zeta)$ is the Bessel function of the first kind. Suppose the electrical conductivity of the substrate is much larger than that of the cladding. In this case, a plot of R_{cs} as a function of h_c/r, shown in Figure 17.9, describes the dependence of R_{cs} on h. It is observed that a fully developed Maxwell constriction resistance requires that $h_c > 10r$. For $h_c < 10r$, one gets that $R_{cs} < R_M$, an important consideration when calculating electrical and thermal point contact resistances.

Consider now the same case as before but here the substrate is Pt. The constriction resistance, R_{cs}, as a function of h_c/r, shown in Figure 17.10 by the solid line, will continuously change from the value associated with a Pt medium (dashed line) to that of an Al medium (dotted line).

17.3.2
Electrical Resistance in the Sharvin Limit

We now consider the same geometry as that of Figure 17.1, except that now the mean free path, ℓ, is much larger than the radius of the constriction. The electrical constriction resistance for this case was obtained by Sharvin from the difference between the number of particles striking it from each medium. This resistance, denoted by R_S, is given in terms of ℓ, σ, and r, by

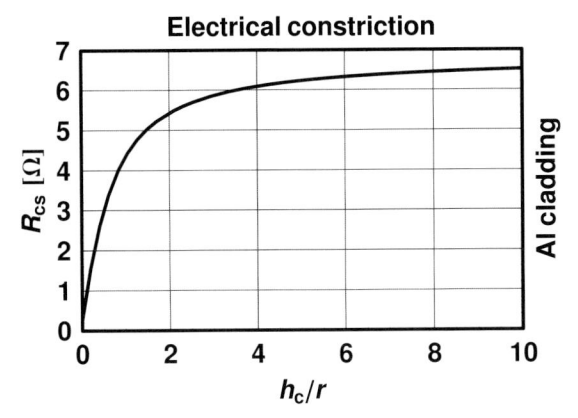

Fig. 17.9 The electrical resistance, R_{cs}, as a function of h_c/r. Here, a perfectly conducting disk with radius r is contacting a metallic cladding film with thickness h_c and electrical conductivity σ_c, which is deposited on a semi-infinite substrate with electrical conductivity $\sigma_s \gg \sigma_c$.

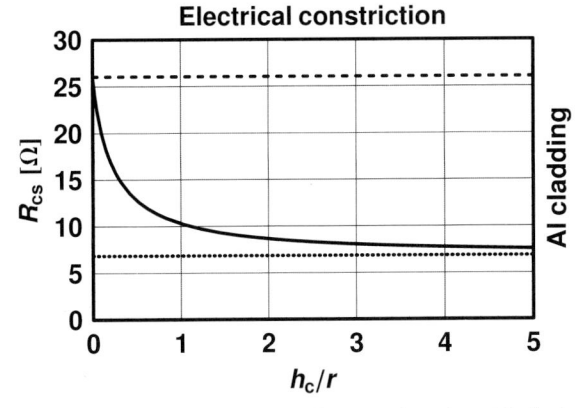

Fig. 17.10 The electrical resistance, R_{cs}, as a function of h_c/r, for an Al cladding film on a Pt substrate (solid line). The dashed and dotted lines refer to the electrical resistance of Pt and Al, respectively.

$$R_S = \frac{4\ell}{3\sigma}\frac{1}{\pi r^2} \tag{17.22}$$

for identical bounding media. As for the Maxwell regime, here too one replaces σ with κ for the case where thermal energy is mediated by phonons, and ℓ refers to the phonons mean free path. Note that for the metallic case, (a) the electrons cross the constriction ballistically, and (b) R_S is inversely proportional to the area, rather than to the radius of the constriction. It is interesting to note that R_S has the same value as that of a cylindrical resistor with radius r and length $4\ell/3$, and has therefore a constant value for a medium

17.3 Constriction Resistance

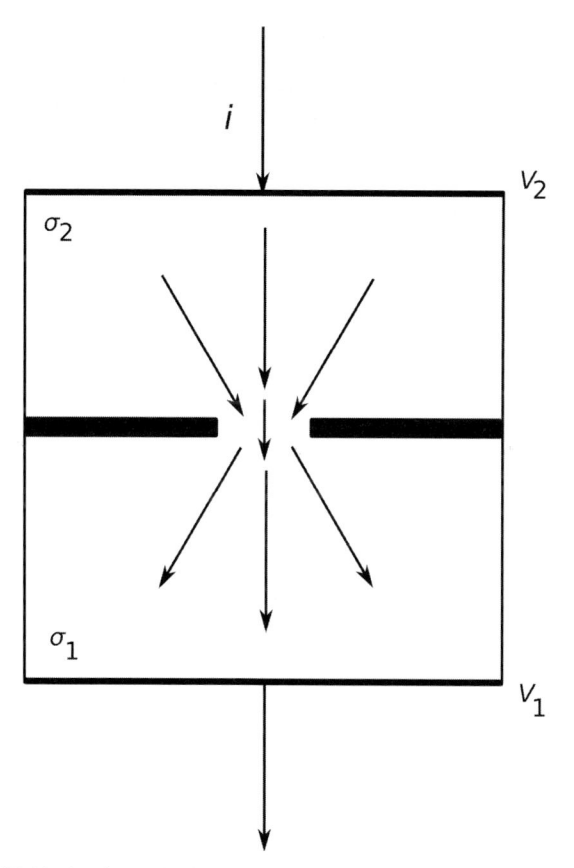

Fig. 17.11 A schematic diagram similar to that of Figure 17.1, except that here the mean free path, ℓ, is much larger than the radius r of the constriction.

whose thickness $h > \ell$. As for the Maxwell regime, here too R_S is independent of the thickness of the two bounding media provided that $h > \ell$, because the major contribution to its value comes from the volume close to the constriction. For dissimilar media, one uses the replacement

$$\frac{\ell}{\sigma} \to \frac{1}{2}\left(\frac{\ell_1}{\sigma_1} + \frac{\ell_2}{\sigma_2}\right). \tag{17.23}$$

The electrical conductance per unit area, G_S, in terms of ℓ and σ, is given by

$$G_S = \frac{3}{4}\frac{\sigma}{\ell} \tag{17.24}$$

for the same bounding media, while for dissimilar media one uses the replacement

$$\frac{\sigma}{\ell} \to \frac{2}{\frac{\ell_1}{\sigma_1} + \frac{\ell_2}{\sigma_2}}.\tag{17.25}$$

The electrical resistance, in terms of $\mathcal{E}_F$ and r, is

$$R_S = \frac{1}{4\pi}\frac{h^3}{e^2 m_e}\frac{1}{e\mathcal{E}_F}\frac{1}{\pi r^2}\tag{17.26}$$

for the same bounding media, while for dissimilar media one uses the replacement

$$\mathcal{E}_F \to \frac{2}{1/\mathcal{E}_{F_1} + 1/\mathcal{E}_{F_1}}.\tag{17.27}$$

G_s, in terms of $\mathcal{E}_F$ is

$$G_S = 4\pi\frac{e^2 m_e}{h^3}e\mathcal{E}_F \tag{17.28}$$

for the same bounding media, while for dissimilar media one uses the replacement given in Eq. (17.27). Figure 17.12 shows the Sharvin spreading resistance, R_S, as a function of the radius of the constriction, r, for identical bounding media. Here, the solid, dashed, and dotted lines refer to Cu, Al, and Pt media, respectively.

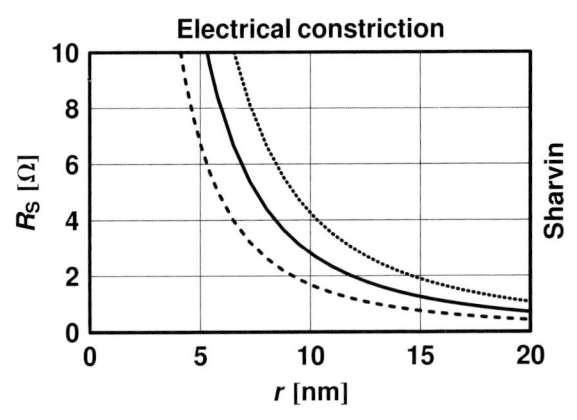

Fig. 17.12 The Sharvin spreading resistance, R_S, as a function of the radius of the constriction, r. Here, the solid, dashed, and dotted lines refer to Cu, Al, and Pt media, respectively.

17.3.3
Combined Electrical Resistance

The solution to the spreading resistance due to a constriction for any value of K, denoted by R_c, can be approximated by the scaling function, γ_c, given by

$$\gamma_c = \frac{1 + 0.83\ell/r}{1 + 1.33\ell/r}.\tag{17.29}$$

Figure 17.13 shows γ_c for the range $0 \ll \ell/r \ll 10$. For $\ell/r = 0$ and for $\ell/r = \infty$ one obtains $\gamma_c = 1$ and 0.626, respectively.

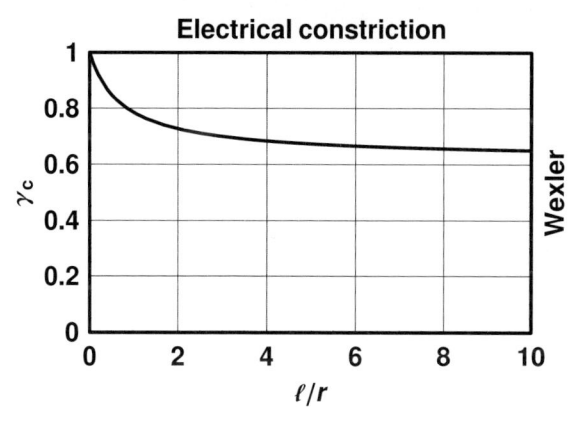

Fig. 17.13 The function γ_c as a function of ℓ/r.

The spreading resistance, R, using γ_c, is given by

$$R = \gamma_c R_M + R_S. \tag{17.30}$$

R_c, in terms of ℓ, σ and r, is

$$R = \frac{\gamma_e}{\sigma}\frac{1}{2r} + \frac{4\ell}{3\sigma}\frac{1}{\pi r^2} \tag{17.31}$$

for the same bonding media, while for dissimilar media one uses the replacements

$$\frac{\gamma_e}{\sigma} \rightarrow \frac{1}{2}\left(\frac{\gamma_{e_1}}{\sigma_1} + \frac{\gamma_{e_2}}{\sigma_2}\right) \tag{17.32}$$

and Eq. (17.23). R in terms of ℓ, $\mathcal{E}_F$, and r is given by

$$R = \frac{1}{4\pi}\frac{h^3}{e^2 m_e}\left(\frac{3}{4}\frac{\gamma_e}{e\mathcal{E}_F\ell}\frac{1}{2r} + \frac{1}{e\mathcal{E}_F}\frac{1}{\pi r^2}\right) \tag{17.33}$$

for the same bounding media, while for dissimilar media one uses the replacement

$$\frac{\gamma_e}{\mathcal{E}_F\ell} \rightarrow \frac{1}{2}\left(\frac{\gamma_{e_1}}{\mathcal{E}_{F_1}\ell_1} + \frac{\gamma_{e_2}}{\mathcal{E}_{F_2}\ell_2}\right). \tag{17.34}$$

Figure 17.14 shows the constriction resistance, R, as a function of the radius of the constriction, r, for identical bounding media. Here, the solid, dashed, and dotted lines refer to Cu, Al, and Pt media, respectively.

The radius at which the contributions from the scaled Maxwell resistance equals the Sharvin resistance, r_c, is given in Table 17.2.

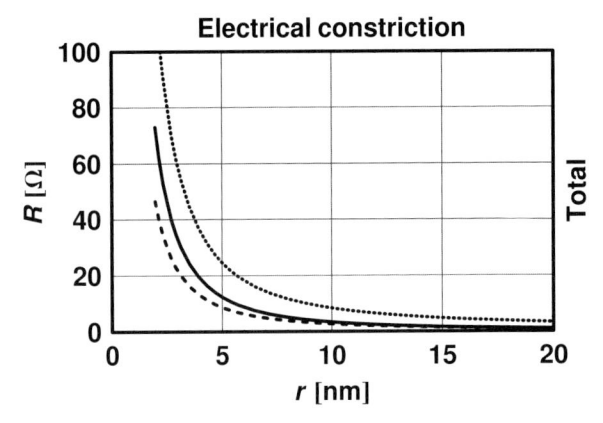

Fig. 17.14 The constriction resistance, R, as a function of the radius of the constriction, r, for identical bounding media. Here, the solid, dashed, and dotted lines refer to Cu, Al, and Pt media, respectively.

Table 17.2 The radius r_c at which the contributions from the scaled Maxwell and the Sharvin resistances equal each other.

	Pt	Al	Cu
r_c (nm)	10.3008	15.4985	41.7313

17.4 Boundary Resistance

17.4.1 Thermal Boundary Resistance of General Media

We will now treat a case in which a temperature gradient is applied between two bounding semi-infinite media, as shown in Figure 17.15. These two media are separated by a thin boundary that has no constriction but rather acts diffusely to scatter phonons or electrons incident on it in all directions, including the adjacent medium. In the following, we use the diffuse mismatch model (DMM), first developed for phonons, with a temperature-consistent definition. Here, the equivalent temperature of all phonons on each side is considered. The transmissivity, $T_{i,j}$, where i and j refer to each side of the semi-infinite media, is given by

$$T_{ij} = \frac{C_j v_j}{C_1 v_1 + C_2 v_2}. \tag{17.35}$$

where C_j and v_j denote the heat capacity per unit volume and group velocity of the phonons belonging to each medium, respectively. The thermal boundary resistance, $\mathcal{R}_b$, for a given area $A = \pi r^2$, in terms of T_{ij}, is given by

17.4 Boundary Resistance

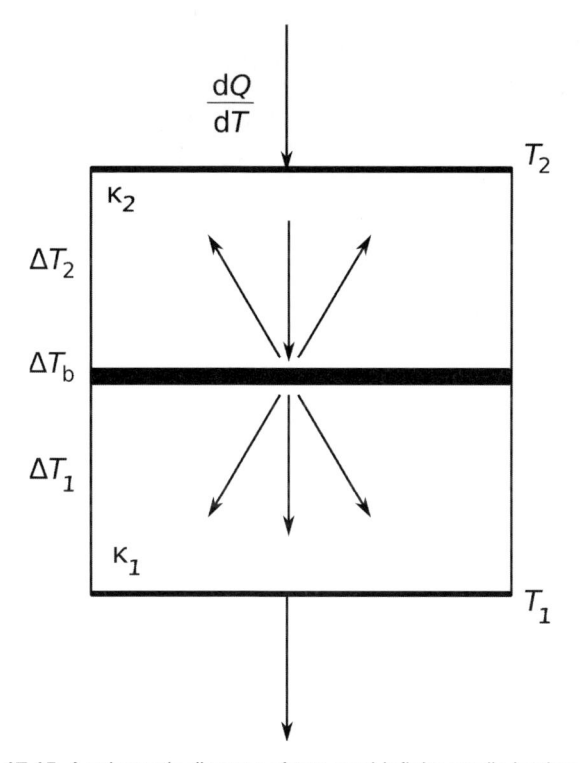

Fig. 17.15 A schematic diagram of two semi-infinite media having thermal conductivities κ_1 and κ_2, separated by a diffusing boundary (denoted by a thick solid line) with thickness h. The two media are held at temperatures T_1 and T_2, resulting in a heat flow dQ/dt. The direction and length of the arrows indicate the direction of the flow of the electrons and the magnitude of their mean free path, respectively. Here, ΔT_1, ΔT_2, and ΔT_B refer to the temperature drop across the top and bottom media, and across the diffusing boundary, respectively.

$$\mathcal{R}_b = 4 \frac{1 - \frac{1}{2}(T_{12} + T_{21})}{T_{21} C_2 v_2} \frac{1}{\pi r^2}. \tag{17.36}$$

Inserting the value of T_{ij} in $\mathcal{R}_b$ gives

$$\mathcal{R}_b = 2 \frac{C_1 v_1 + C_2 v_2}{C_1 v_1 C_2 v_2} \frac{1}{\pi r^2}. \tag{17.37}$$

Note that the classical expression given by the diffuse mismatch model yields a value that is twice that of Eq. (17.37). However, one should employ Eq. (17.37) for the case in which heat conductivity of two media separated by a diffusing boundary is calculated, because it takes into account the totality of the diffused phonons. Using the expression for the heat conductivity, κ_i,

$$\kappa_i = \frac{1}{3} C_i v_i \ell_i. \tag{17.38}$$

yields the final expression for the thermal boundary resistance for identical media in terms of ℓ, κ, and r,

$$R_b = \frac{4}{3}\frac{\ell}{\kappa}\frac{1}{\pi r^2}. \tag{17.39}$$

For dissimilar media, one uses the replacement

$$\frac{\ell}{\kappa} \to \frac{1}{2}\left(\frac{\ell_1}{\kappa_1} + \frac{\ell_2}{\kappa_2}\right). \tag{17.40}$$

17.4.2
Thermal Boundary Resistance of Metallic Media

From now on, we consider the thermal and electrical resistance of two semi-infinite metallic media separated by a diffusing boundary, for which thermal energy is mediated by electrons. The thermal boundary resistance for identical media, in terms of T, $\mathcal{E}_F$ and r, is given by

$$R_b = \frac{3}{4\pi^3}\frac{h^3}{m_e K_B^2 T}\frac{1}{e\mathcal{E}_F}\frac{1}{\pi r^2}. \tag{17.41}$$

For dissimilar media, one uses the replacement given in Eq. (17.27). The thermal boundary conductance per unit area, $\mathcal{G}_b$, for identical media, in terms of ℓ and κ, is

$$\mathcal{G}_b = \frac{3}{4}\frac{\kappa}{\ell}. \tag{17.42}$$

For dissimilar media, one uses the replacement

$$\frac{\kappa}{\ell} \to \frac{2}{\frac{\ell_1}{\kappa_1} + \frac{\ell_2}{\kappa_2}}. \tag{17.43}$$

The thermal boundary conductance per unit area, $\mathcal{G}_b$, for identical media, in terms of T and $\mathcal{E}_F$, is

$$\mathcal{G}_b = \frac{4\pi^3}{3}\frac{m_e K_B^2 T}{h^3} e\mathcal{E}_F. \tag{17.44}$$

For dissimilar media, one uses the replacement given in Eq. (17.27). Figure 17.16 shows the boundary resistance, R_b, as a function of r for identical bounding media. Here, the solid, dashed, and dotted lines refer to Cu, Al, and Pt media, respectively.

One can also define an admittance per unit area for a degenerate electron gas, $\mathcal{Y}$, by

$$\mathcal{Y} = C_e v_F, \tag{17.45}$$

17.4 Boundary Resistance

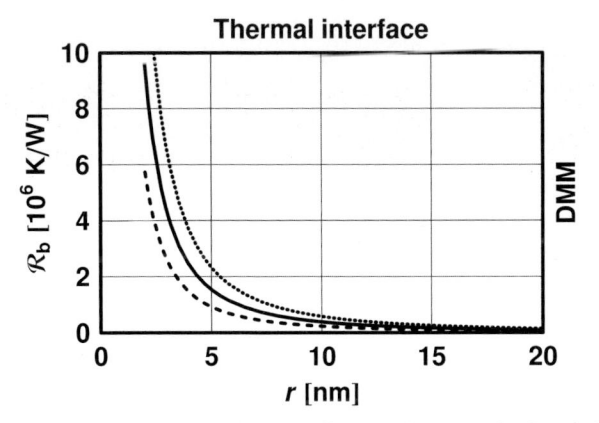

Fig. 17.16 The boundary resistance, $\mathcal{R}_b$, as a function of r. Here, the solid, dashed, and dotted lines refer to Cu, Al, and Pt media, respectively.

where v_F and C_e are given by Eq. (17.3) and Eq. (17.13), respectively. The resulting expression for $\mathcal{Y}$,

$$\mathcal{Y} = \frac{16\pi^3}{3} \frac{m_e K_B^2 T}{h^3} e\mathcal{E}_F, \qquad (17.46)$$

is shown in Figure 17.17.

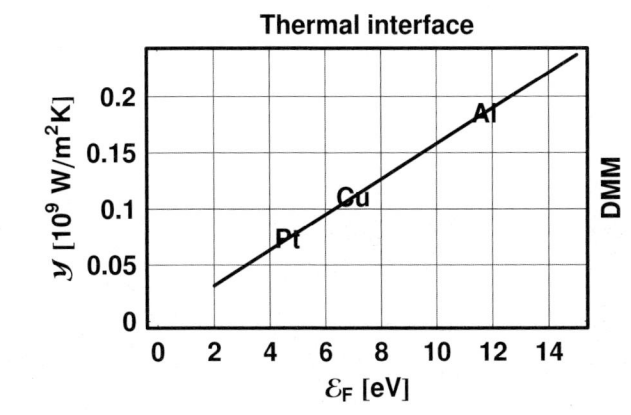

Fig. 17.17 The admittance per unit area, $\mathcal{Y}$, as a function of the Fermi energy, $\mathcal{E}_F$.

The thermal boundary conductance per unit area, $\mathcal{G}_b$, for identical media, in terms of $\mathcal{Y}$, is given in the diffuse mismatch model by

$$\mathcal{G}_b = \frac{1}{4}\mathcal{Y}. \qquad (17.47)$$

For dissimilar media, in accordance with Eq. (17.37), one uses the replacement

$$\mathcal{Y} \to \frac{2}{1/\mathcal{Y}_1 + 1/\mathcal{Y}_2}. \tag{17.48}$$

The thermal boundary conductance per unit area, $\mathcal{G}_b$, for identical media, in terms of T and $\mathcal{E}_F$, using Eq. (17.46) and Eq. (17.47), is identical to that given by Eq. (17.44). Figure 17.18 shows the conductance per area, $\mathcal{G}_b$, as a function of $\mathcal{E}_F$, for identical bounding media.

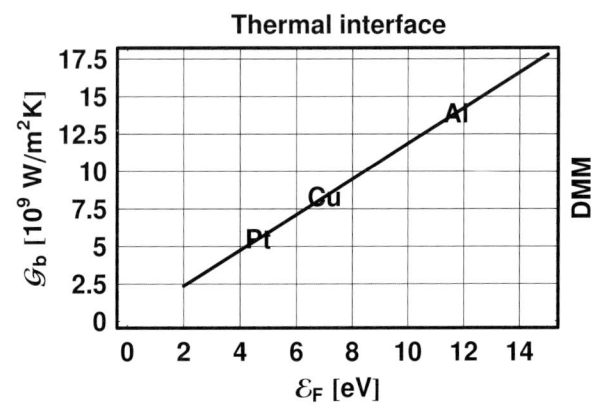

Fig. 17.18 The conductance per area, $\mathcal{G}_b$, as a function of the Fermi energy, $\mathcal{E}_F$.

17.4.3
Electrical Boundary Resistance of Metallic Media

It is useful to note that to transform metallic thermal resistance, $\mathcal{R}$, to electrical resistance, R, one uses

$$R = \mathcal{R} T N_{\text{Lorenz}}, \tag{17.49}$$

yielding

$$R_b = R_S = \frac{1}{4\pi} \frac{h^3}{e^2 m_e} \frac{1}{e \mathcal{E}_F} \frac{1}{\pi r^2}. \tag{17.50}$$

To transform metallic thermal conductance per unit area, $\mathcal{G}$, to electrical conductance per unit area, G, one uses

$$G = \mathcal{G}/(T N_{\text{Lorenz}}), \tag{17.51}$$

yielding

$$G_b = G_s = 4\pi \frac{e^2 m_e}{h^3} e\mathcal{E}_F. \tag{17.52}$$

We find, therefore, that for metals, the equations for the constriction resistance and for the conductance per unit area for $\ell \gg r$ are identical to those of a boundary under diffuse mismatch conditions. Note that the Sharvin and boundary resistances do not include the contributions to the resistance from the bounding media.

The modeling presented in this chapter is currently a hot topic that is being investigated by several groups both theoretically and experimentally. The references given at the end of this chapter are only a small selection from a large body of literature covering these topics. As a summary, Table 17.3 gives the numerical value of all the properties calculated in this chapter for Cu, Al, and Pt. These values are close to those found in the literature; that may vary somewhat depending on the quality of the probed media and the experimental methods used in the measurements.

Table 17.3 The numerical values of all the properties calculated in this chapter for Cu, Al, and Pt.

	Cu	Al	Pt	Units
σ	58.8	36.5	9.6	10^6 $1/m\Omega$
κ	420.896	261.27	68.7177	W/mK
$\mathcal{E}_F$	7	11.7	4.63	eV
v_F	1.569 19	2.0287	1.276 19	10^6 m/s
T_F	81.2311	135.772	53.7286	10^3 K
N_e	8.410 88	18.1749	4.524 45	10^{28} $1/m^3$
k_F	13.5546	17.5239	11.0237	10^9 $1/m$
ℓ	38.9295	14.4579	9.609 26	nm
C_e	70.5464 T	91.205 T	57.3741 T	$J/m^3 K$
y	0.1107 T	0.185 028 T	0.073 220 4 T	10^9 $W/m^2 K$
G_b	0.027 675 1 T	0.046 257 T	0.018 305 1 T	10^9 $W/m^2 K$
G_b	1.132 82	1.893 42	0.749 277	10^{15} $W/m^2 K$

▶ **Exercises for Chapter 17**

1. What is the physical nature of R_S and R_b?
2. In the presence of both a Sharvin-type restriction and a DMM-type boundary, does one add their individual contributions or are they one and the same?
3. What role do the thicknesses of the constriction and the boundary play?

4. What exactly is the nature of a DMM-type boundary?

References

1. G. Wexler, "The size effect and the non-local Boltzmann transport equation in the orifice and disk geometry," Proc. Phys. Soc. **89**, 927 (1966).
2. B. Nikolic and P. B. Allen, "Electron transport through a circular constriction," Phys. Rev. B **60**, 3963 (1999).
3. M. M. Yovanovich, J. R. Culham and P. Teertstra, "Analytical modeling of spreading resistance in flux tubes, half spaces, and compound disks," IEEE Trans. Components, Packaging, and Manufacturing Technology, **21**, 158 (1998).
4. G. Chen, "Thermal conductivity and ballistic-phonon transport in cross-plane direction supperlattices," Phys. Rev. B **57**, 14958 (1998-I).
5. B. C. Gundrum, D. G. Cahill, and R. S. Averback, "Thermal conductance of metal-metal interfaces," Phys. Rev. B **72**, 245426 (2005).
6. R. Prasher, "Predicting the thermal resistance of nanosized constrictions," Nano Lett. **5**, 2155 (2005).

18
Scanning Thermal Conductivity Microscopy

■ Highlights

1. Thermal resistance of a cantilever, tip, sample, and tip–sample interface
2. Mechanical and thermal bending of a cantilever
3. Theory of operation of scanning thermal conductivity microscopy

Abstract

This chapter describes an implementation of an atomic force microscope (AFM) that can map thermal conductivity features across a sample with a high spatial resolution. The microscope employs a single-side, metal-coated cantilever, which acts as a bimetallic strip, together with a heating laser whose beam is focused on the cantilever's free end, on the opposite side of its tip. Subtracting the topography obtained by the unheated and heated cantilever yields a map of thermal conductivity across the surface of a sample. The chapter presents the theory of operation of the microscope that was found to be in agreement with experimental results obtained on a silicon sample patterned with oxide features.

18.1
Introduction

Local thermal properties of materials play an important role in many applications where characterizing heat dissipation from nanostructures is essential. Consider, for example, the semiconductor industry which is currently experiencing a need for an increasing density and speed of operation of transistors in processors and solid-state storage chips. Heat dissipation is becoming a limiting factor as these devices shrink in size, requiring better tools that are able to characterize thermal conductivity of nanoscale structures. As a further example, consider the next generation of hard disk drives, whose bits are expected to shrink to such a small size that their magnetic properties approach the super-paramagnetic limit. Overcoming this limit requires a dramatic increase in the coercivity of their magnetic material, which in turn will make them resistant to switching by conventional magnetic heads. There is currently an intense research effort to reduce the coercivity of these bits by heating them during the write process using near-field optics. Understanding the thermal response of nanoscale electronic devices and magnetic bits is therefore of scientific and technical importance, requiring new approaches to thermal conductivity measurements.

Characterization of thermal properties of a sample has been actively pursued using a variety of interactions that can take place between the apex of a tip of a cantilever and the surface of a sample. Examples of such tip–sample interactions include (a) temperature-dependent contact potential difference (CPD), (b) Raman scattering, (c) temperature dependence of the reflectivity of a cantilever, (d) bimetal effect that bends a one-side, metal-coated cantilever, and (e) a submicron thermocouple fabricated on the apex of a tip. Among these interactions, the most sensitive one is the bimetal effect that can yield a bending sensitivity of up to 10 nm/µW of power injected into the free end of a one-side, metal-coated cantilever.

This chapter presents a novel implementation of this effect using an AFM that yields thermal conductivity features across the surface of a sample. This microscope, dubbed κ-AFM and shown schematically in Figure 18.1, employs such a cantilever together with a heating laser that enables one to scan the same area on a sample with an unheated and heated cantilever. As will be discussed later, while the low-power monitoring laser of the AFM is focused somewhere along the cantilever, the high-power heating laser is focused on the free end of the cantilever, opposite the tip side. In (a), the tip of the cantilever is not in contact with the sample, and the HeNe laser beam is focused on the free end of the cantilever. The absorbed light heats up the cantilever and bends it. In (b), the tip makes contact with a feature whose thermal conductivity is high. As a result, the cantilever cools down and its curvature is reduced. Conversely, when the tip contacts a feature whose thermal conductivity is low, the curvature of the cantilever increases, as shown in (c). The κ-AFM, like any conventional AFM, does not measure the position of the free end of a bent cantilever, but rather the slope of the cantilever at the point where the monitoring laser is focused. This slope is obtained from the differential photocurrent generated by a quadrant photodiode, upon which the reflected beam of the monitoring laser is incident. The role of the electronic feedback (EFB) of an AFM is to maintain this differential photocurrent constant at a predetermined value as the tip raster scans across a sample. The EFB accomplishes this task by changing the position of the base of the cantilever, relative to that of the sample, using a piezotube or a voice coil. Operating the AFM with feedback, therefore, keeps the bending of the cantilever constant during the imaging process. The topography of the sample is obtained from the signal that drives the piezotube or voice coil, which reflects the local height of the sample under the tip.

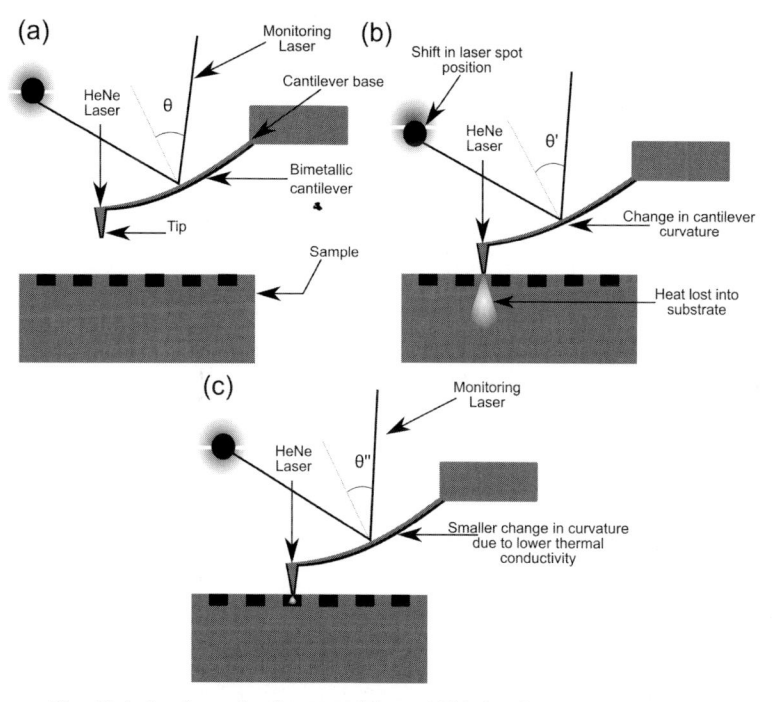

Fig. 18.1 A schematic diagram of the κ-AFM showing a conventional SPM equipped with a one-side, metal-coated cantilever and an additional heating laser incident on the free end of the cantilever, on the opposite side of its tip.

Consider now the effect that the heating laser has on the operation of the κ-AFM. The heating laser will inject a constant thermal power, $P = P_{in}$, into the free end of the cantilever whose tip is in good thermal contact with a sample. Part of the thermal power will flow from the free end of the cantilever into its base, while the rest will flow through the tip into the sample. The circuit that describes the thermal flow, shown in Figure 18.2, consists of the thermal resistors corresponding to the cantilever, R_c, the tip, R_t, and the sample, R_s. Note that R_c is in parallel with R_t which is in series with R_s. Because a constant thermal power feeds the circuit during the operation with the heated cantilever, changes in R_s will change the total thermal resistance of the circuit, modifying the temperature of the cantilever. This change in temperature modifies the cantilever slope, thus distorting the observed local topographic features relative to those observed with the unheated cantilever. This distortion in topography is a function of the local thermal conductivity of the sample. In a typical use of the κ-AFM, a sample is imaged twice, first with the unheated and next with the heated cantilever. As will be explained in detail in the following sections, the difference between these two images yields a map of thermal conductivity features across the surface of the sample.

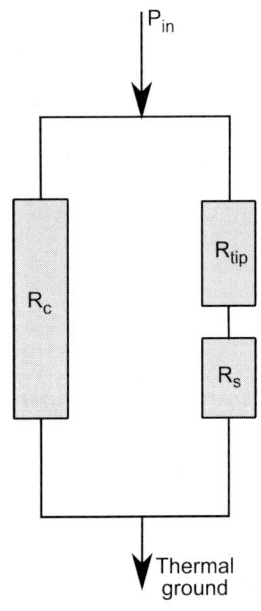

Fig. 18.2 A schematic diagram showing the injected thermal power, P_{in}, flowing into the thermal resistors comprising the cantilever, R_c, in parallel with that of the tip, $R_t = R_{tip}$, which is in series with that of the tip–sample point contact, R_s.

The notations relating to the geometry of the cantilever consist of its length, ℓ_c, width, w_c, and thickness, h_c. For the tip, the notations consist of its height, h_t, the half angle of the cone that models its shape, θ_t, and its apex radius, r_t. The numerical values of these parameters, used in the examples, are given in Table 18.1 and Table 18.2.

Table 18.1 The cantilever length, ℓ_c, width, w_c, and thickness, h_c.

$\ell_c = 444\,\mu m$
$w_c = 51\,\mu m$
$h_c = 1.9\,\mu m$

Table 10.2 The tip height, h_t, the half angle of the cone that models its shape, θ_t, and its apex radius, r_t.

$$h_t = 17.5\,\mu m$$
$$\theta_t = 15\text{ deg}$$
$$r_t = 30\,nm$$

18.2 Theory of Thermal Response

18.2.1 Electrical and Thermal Circuits

Before embarking on an analysis of the thermal circuit depicted in Figure 18.2, it is worthwhile to recount the analogy between electric and thermal circuits. The self-explanatory items in Table 18.3 denote length, area, and volume by ℓ, A, and V, respectively, and mass density and specific heat by ρ_m and C_p, respectively.

Table 18.3 The analogy between electrical and thermal circuits. The self-explanatory items denote length, area, and volume by ℓ, A, and V, respectively, and mass density and specific heat by ρ_m and C_p, respectively.

Electric		Thermal	
Charge	Q_e	Heat	Q_{th}
Current	i	Laser power	$P_{th} = \frac{\partial Q}{\partial t}$
Potential	v	Temperature	T
Resistivity	ρ_e	Resistivity	ρ_{th}
Resistance	$R_e = \rho_e \frac{\ell}{A}$	Resistance	$R_{th} = \rho_{th} \frac{\ell}{A}$
Capacitance	$C_e = \frac{\partial Q_e}{\partial v}$	Heat capacity	$C_{th} = \frac{\partial Q_{th}}{\partial T}$

Analysis of the thermal circuit shown in Figure 18.2, considering the analogy between electric and thermal circuits, leads to the conclusion that the sensitivity of the κ-AFM to thermal conductivity features across a sample will be optimized by employing a cantilever–tip system such that the ratio of R_t/R_c is made as large as practically possible, while maintaining the ratio R_t/R_s as small as possible. The reason for this choice is that a large portion of the thermal power injected into the free end of the cantilever should be diverted through the tip into the sample, rather than into its base. For a cantilever–tip system made of the same material, a preferred choice is a long, thin and narrow structure made of silicon. Silicon has a large, albeit temperature dependent thermal conductivity, κ_{Si}, that strikes a good balance between R_c, R_t, and R_s. To calculate the thermal response of the cantilever, tip, and sample, it is necessary to use the temperature dependent thermal resistivity of silicon, $\rho_{Si} = 1/\kappa_{Si}$, given by $\rho_{Si} = \rho_0 + \rho_T \Delta T$, where $\rho_0 = 194/29\,900$ m K/W, $\rho_T = 1/29\,900$ m/W, and ΔT is the temperature above ambient (20 °C).

18.2.2
Cantilever Thermal Resistance and Temperature

The temperature, ΔT_c, along a laser-heated cantilever, whose tip is not in contact with a sample, at a point z away from its base, is obtained from the solution of

$$\frac{\partial}{\partial z}\Delta T_c = P\frac{\rho_{Si}}{w_c h_c} \tag{18.1}$$

given by

$$\Delta T_c = \frac{\rho_0}{\rho_T}\left(e^{P\frac{\rho_T}{h_c w_c}z} - 1\right). \tag{18.2}$$

Figure 18.3 shows the temperature across the cantilever, ΔT_c, as a function of the injected power, P, where the solid and dashed lines refer to the inclusion or exclusion of the temperature dependence of the thermal resistivity of silicon, respectively. As observed in Figure 18.3, it is important to include the temperature dependence for $P > 1$ mW.

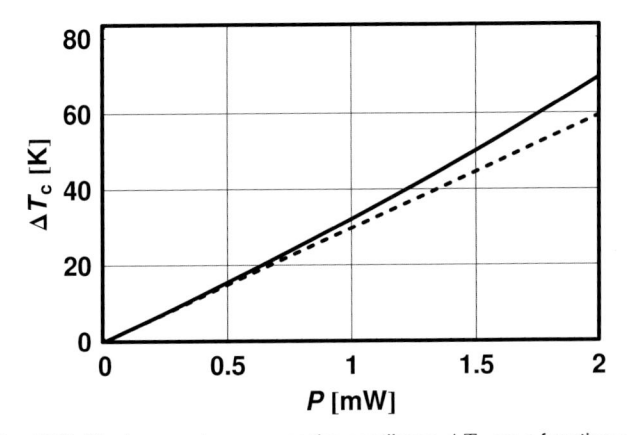

Fig. 18.3 The temperature across the cantilever, ΔT_c, as a function of the injected thermal power, P. The solid and dashed lines refer to the inclusion or exclusion of the temperature dependence of the thermal resistivity of silicon, respectively.

Figure 18.4 shows the temperature along the cantilever, ΔT_c, as a function of z for $P = 0.1$ mW (solid line), 0.5 mW (dashed line), and 2 mW (dotted line).

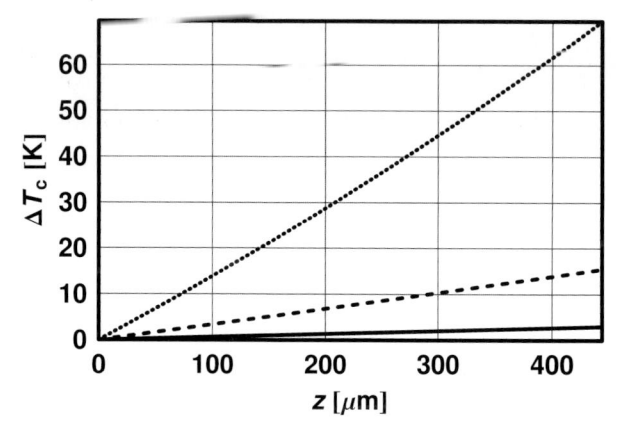

Fig. 18.4 The temperature along the cantilever, ΔT_c, as a function of z, for $P = 0.1\,\text{mW}$ (solid line), $0.5\,\text{mW}$ (dashed line), and $2\,\text{mW}$ (dotted line). Here the cantilever base is at $z = 0\,\mu\text{m}$, and $z = 444\,\mu\text{m}$ corresponds to the position of the tip.

Note that the base of the cantilever is at $z = 0$, and that for moderate values of P, the curves in Figure 18.3 and Figure 18.4 are almost linear.

18.2.3
Tip–Sample Thermal Resistance

The tip–sample contact area serves as a constriction through which thermal energy flows from a heated tip into the sample. The thermal resistance due to this constriction, R_s, is the sum of the Maxwell and Sharvin resistances (see Chapter 17), given by

$$R_s = \frac{1}{\kappa}\frac{1}{4r_t}\left(\gamma_c + \frac{16}{3\pi}\frac{\ell_{ph}}{r_t}\right), \tag{18.3}$$

where ℓ_{ph} is the mean free path of the phonons (mfp), and γ_c is a scaling function given by

$$\gamma_c = \frac{1 + 0.83\ell_{ph}/r_t}{1 + 1.33\ell_{ph}/r_t}. \tag{18.4}$$

Note that r_t is an effective tip–sample thermal contact radius determined by the apex of the bare tip, the metal coating, and by the surrounding liquid meniscus. In the case presented here, the effective tip radius is taken to be 30 nm, which is the sum of its radius of curvature ($\sim 10\,\text{nm}$) and its metal coating ($\sim 20\,\text{nm}$).

18.2.4
Tip Thermal Resistance and Temperature

The tip geometry can be modeled as that of a truncated silicon cone whose thermal resistivity, ρ_t, is temperature dependent. The temperature along the tip, ΔT_t, at a height h above its apex, can be obtained from the solution of the differential equation

$$\frac{\partial}{\partial h}\Delta T_t = P_t \rho_t \frac{1}{\pi(r_t + h\tan\theta)^2}. \tag{18.5}$$

Here, $h = 0$ is the position of the apex of the tip, $h = h_t$ is the position of the base of the tip and P_t is the power injected into the tip. Note that the temperature of the tip apex is assumed to be equal to that at the top of the sample, leading to the boundary condition

$$\Delta T_t|_{h=0} = P_t R_s. \tag{18.6}$$

We will discuss this assumption in the context of interface thermal resistance at the end of this chapter. Figure 18.5 shows the linear (dashed line) and nonlinear (solid line) temperature across the silicon tip whose apex is in contact with a silicon sample, $\Delta T_t + \Delta T_s$, as a function of the injected power into the tip, P_t.

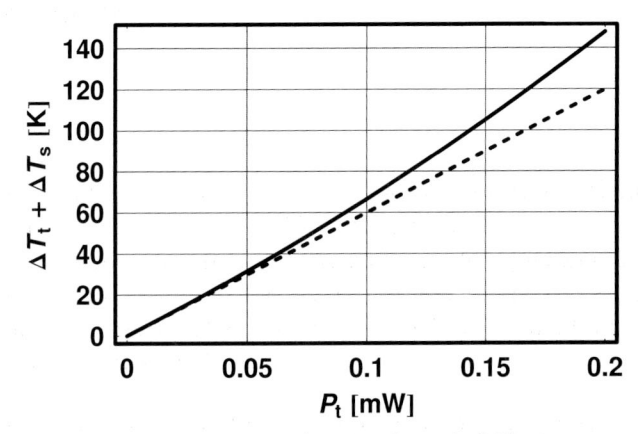

Fig. 18.5 The linear (dashed line) and nonlinear (solid line) temperature across the tip and sample, $\Delta T_t + \Delta T_s$, as a function of the power injected into the tip, P_t.

Figure 18.6 shows the linear (dashed line) and nonlinear (solid line) temperature along the tip height, h, for $P_t = 0.02\,\text{mW}$, where the apex of the tip is at $h = 0$. Note that most of the temperature drops in the vicinity of the apex of the tip. At ambient temperatures, Eq. (18.2) yields the linear thermal resistance of the tip, R_t,

$$R_t = \rho_t \frac{1}{\pi r_t} \frac{1}{\tan\theta + \frac{r_t}{h_t}}, \qquad (18.7)$$

which, for $r_t \ll h_t$ and $\theta = 51.854°$, reverts to the constriction resistance in the Maxwell limit (see Chapter 17). The solution to the thermal circuit presented in Figure 18.2 can be obtained by assuming that the temperature across the cantilever equals the sum of the temperatures across the tip and the sample.

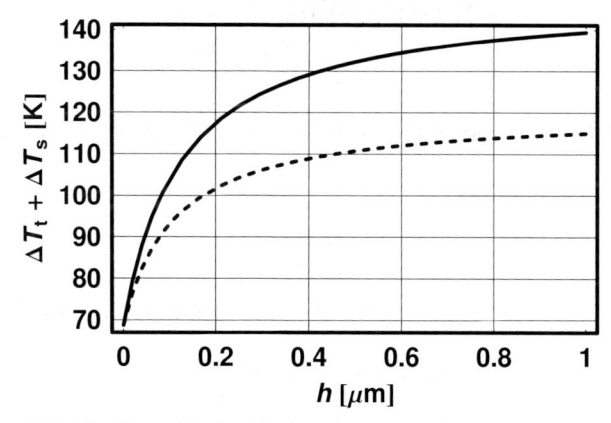

Fig. 18.6 The linear (dashed line) and nonlinear (solid line) temperature along the tip height, h, for $P_t = 0.02\,\mathrm{mW}$, where the apex of the tip is at $h = 0$.

Figure 18.7 shows the temperature across the cantilever, ΔT_c, as a function of the thermal conductivity of the sample for an injected thermal power of 1.68 mW. The solid and dashed lines refer to the cases where the thermal conductivity of silicon is temperature dependent and independent, respectively. One observes (a) the importance of incorporating the temperature dependence of the thermal conductivity of silicon into the solution of the thermal circuit, and (b) that the range of thermal conductivity values in Figure 18.7 covers most media of practical interest.

18.3
Thermal and Mechanical Cantilever Bending

18.3.1
Mechanical Bending

The principle of operation of the κ-AFM is based on the interplay between the mechanically and thermally induced bending of a heated, one-side, metal coated cantilever. The maximum value of the mechanically induced bending

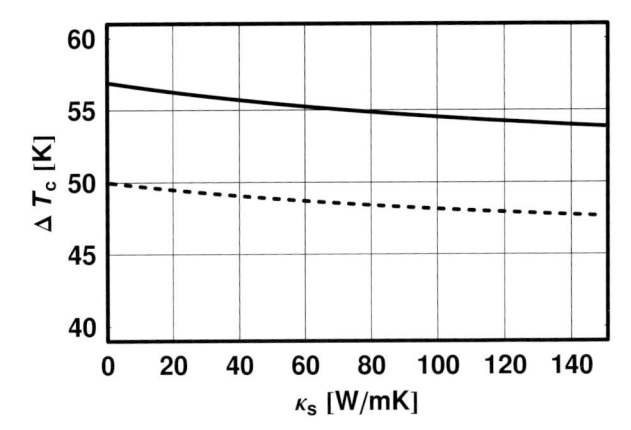

Fig. 18.7 The linear (dashed line) and nonlinear (solid line) solution to the temperature rise, ΔT_c, of a cantilever whose tip is in thermal contact with a sample, as a function of the sample's thermal conductivity, κ_s.

of a cantilever, δy_{max}, derives from the force exerted on its tip, F_{max}, is given in terms of the cantilever spring constant, k, by

$$F_{max} = k\delta y_{max}. \tag{18.8}$$

The mechanically induced bending of the cantilever, δy_m, at a distance z away from its base, is

$$\delta y_m = -\frac{k\delta y_{max}}{6E_c I_r} z^2(z - 3\ell_c). \tag{18.9}$$

Here, E_c and $I_r = w_c h_c^3/12$ are Young's modulus of the cantilever and its moment of inertia, respectively. Differentiating Eq. (18.9) yields the slope associated with the mechanically induced bending, $\delta\theta_m$,

$$\delta\theta_m = -\frac{k\delta y_{max}}{2E_c I_r} z(z - 2\ell_c). \tag{18.10}$$

18.3.2
Thermal Bending

The thermally induced bending of a one-side, metal-coated cantilever, δy_{th}, is obtained from the solution of the differential equation

$$\frac{\partial^2}{\partial z^2}\delta y_{th} = 6(\alpha_f - \alpha_c)\frac{h_f + h_c}{h_f^2 K}\Delta T_c, \tag{18.11}$$

where α_f and α_c are the coefficients of thermal expansion of the metal film and the cantilever, respectively. The constant K is given by

$$K = 4 + 6\frac{h_c}{h_f} + 4\left(\frac{h_c}{h_f}\right)^2 + \frac{E_c}{E_f}\left(\frac{h_c}{h_f}\right)^3 + \frac{E_f}{E_c}\frac{h_f}{h_c}, \qquad (18.12)$$

where E_f and h_f are the Young's modulus and thickness of the metal film coating the cantilever, respectively. A simplified solution of Eq. (18.11) can be written as

$$\delta y_{\text{th, linear}} = P(\alpha_f - \alpha_c)(h_c + h_f)\frac{z^3}{h_f^2 K(\kappa_c h_c + \kappa_f h_f)w_c}, \qquad (18.13)$$

where κ_f and κ_c are the thermal conductivities of the metal and the cantilever, respectively. The derivative of Eq. (18.13), that yields the thermal slope of the cantilever, $\delta\theta_{\text{th, linear}}$, is

$$\delta\theta_{\text{th, linear}} = 3P(\alpha_f - \alpha_c)(h_c + h_f)\frac{z^2}{h_f^2 K(\kappa_c h_c + \kappa_f h_f)w_c}. \qquad (18.14)$$

The mechanically induced bending (solid line) and nonlinear thermally induced bending (dashed line) of the cantilever, for the same total bending, using the parameters given in Table 18.4, are shown in Figure 18.8. Note that the steepest slopes of the mechanically and thermally induced bending are next to the cantilever's base and free end, respectively.

Table 18.4 Thermal expansion of the cantilever and the metal film, α_c and α_f, respectively, Young's modulus of the cantilever, E_c, and the metal film thickness, h_f.

$\alpha_c = 2.6\,\mu\text{m}/\text{mK}$
$\alpha_f = 8.8\,\mu\text{m}/\text{mK}$
$E_c = 179\,\text{GN}/\text{m}^2$
$h_f = 20\,\text{nm}$

18.3.3 Combined Solution

One can now use Eq. (18.8) to Eq. (18.14) to analyze the operation of the κ-AFM in terms of the interplay between the mechanically and thermally induced bending of the cantilever. Consider initially the response of the EFB of the κ-AFM to an unheated cantilever. As the tip is engaged, the cantilever is bent mechanically until it reaches a predetermined set-point dictated by the operator of the system. Figure 18.9 shows the shape of the cantilever before (solid

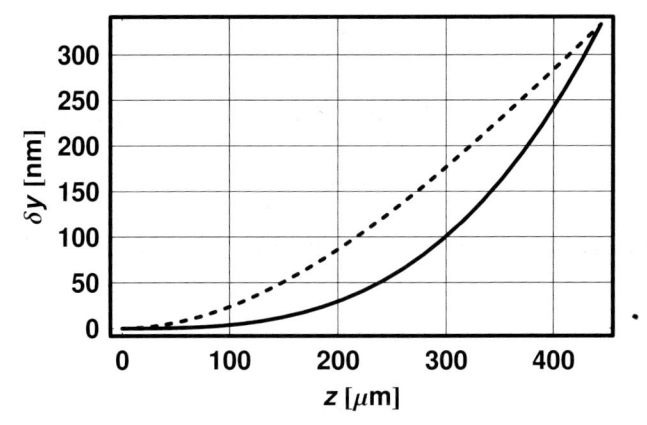

Fig. 18.8 The mechanically induced bending (dashed line) and thermally induced bending (solid line) of the cantilever, δy, for the same value of the total bending, showing the difference in their slopes along the length of the cantilever.

line) and after (dashed line) engagement, where the fixed base of the cantilever is at $z = 0\,\mu\text{m}$, and the positions of the tip for both cases, denoted by V. For simplicity, we chose the position of the sample and tip apex to be both fixed at $\delta y = 0\,\mu\text{m}$. Note that the initial slope of the cantilever depicted in Figure 18.9 is much smaller than when it is mounted in an AFM.

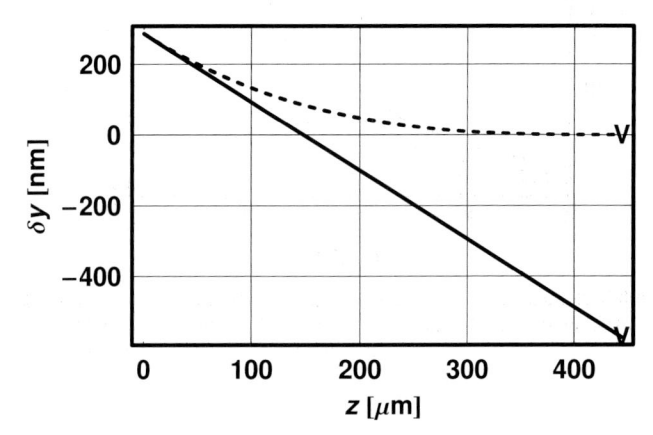

Fig. 18.9 The mechanically induced bending of the free, unheated cantilever, δy, as a function of position z, before (solid line) and after (dashed line) tip–sample engagement. The positions of the tip are denoted by V.

Consider now the operation of the κ-AFM while the laser is heating the cantilever. As the tip of the cantilever scans over regions of varying thermal conductivities, the temperature of the cantilever changes due to the variance in heat transfer into the sample. Thus, the thermal bending of the cantilever changes, which the EFB of the κ-AFM compensates for. This compensation will slightly distort the topography measured by the κ-AFM, as compared to that which is obtained in the unheated case. To understand how this affects the topography measured by the κ-AFM, one has to solve for the combined mechanical and thermal interactions subject to two boundary conditions. The first boundary condition dictates that the tip be in contact with the sample at all times, namely, is fixed at $\delta y = 0$. This condition is satisfied if one initially fixes the EFB set-point at such a value that the cantilever is sufficiently bent by the mechanical force so it can compensate for any subsequent thermally induced bending. The second boundary condition dictates that the slopes of the unheated and heated cantilever, at the point from which the monitoring laser is reflected, are kept at the same value by the EFB. The scaled position of the point where the monitoring laser is focused, is denoted by η. Here, $\eta = 0$ and 1 denote points at the fixed ($z = 0$) and free ends ($z = \ell_c$) of the cantilever, respectively.

The solution to this double boundary condition problem will now be applied to the cases where the tip of the cantilever is in contact with a silicon sample and with its oxide features, whose thermal conductivity κ is 1 mK/W. For simplicity, we assume that both of these are physically at $\delta y = 0$. In reality, however, the silicon sample may have oxide features as high as 1 µm, yet, to a good approximation, the results of the analysis presented here have been found experimentally to be independent of the topographic features of such features.

18.4
Results

18.4.1
Tip-Side Coating, Upward Thermal Bending: Si and SiO$_2$

Figure 18.10 describes the different contributions to the bending of a tip-side coated cantilever where the laser beam is focused halfway along the cantilever, i.e. at $\eta = 0.5$. Shown are the mechanically induced bending of the unheated cantilever (A), the thermally induced (B) and mechanically induced (C) bending components of the heated cantilever and their sum, (D), all for an injected thermal power of 1.68 mW. The solid and dashed lines refer to the silicon and its oxide features, respectively. Note that to maintain a fixed slope at the halfway point, the mechanical bending, when the tip is on the oxide feature, is necessarily large enough to compensate for the thermal bending.

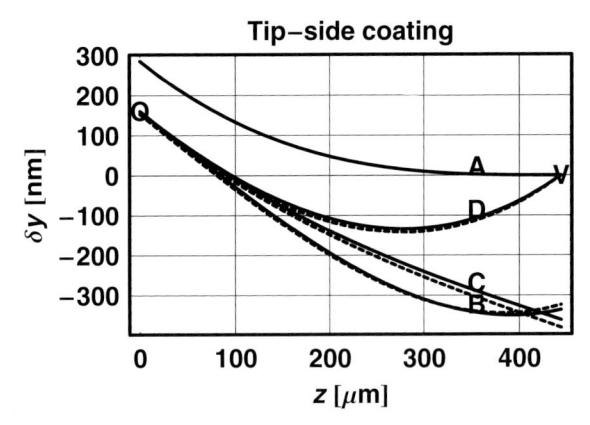

Fig. 18.10 The bending of the unheated, tip-side metal-coated cantilever (A), the thermally induced (B) and mechanically induced (C) bending components of the heated cantilever and their sum, (D), all for an injected thermal power of 1.68 mW. The solid and dashed lines refer to the silicon sample and its oxide features, respectively.

Figure 18.11 is an expanded view of the area denoted by O in Figure 18.10, showing the positions of the base of the cantilever for the silicon sample and its oxide features, denoted by D(Si) and D(SiO$_2$), respectively.

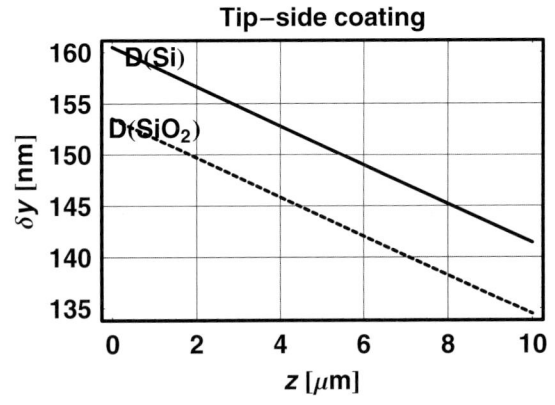

Fig. 18.11 An expanded view of the area denoted by O in Figure 18.10, where the total bending of the heated cantilever, up to 10 μm away from its base, is shown together with the position of the base of the cantilever for the silicon sample and its oxide features, D(Si) and D(SiO$_2$), respectively. Note that the tip is in contact with the sample at $\delta y = 0$.

The most important result obtained in this chapter, namely, that a heated cantilever senses the apparent, rather than the physical height of features having different thermal conductivities, derives from the requirement that the double boundary conditions discussed before is satisfied. This can be accomplished only by having the κ-AFM EFB lower the base of the cantilever as the tip moves from regions of high thermal conductivity to regions of low thermal conductivity. The conclusion of this analysis is that while the oxide features are physically at the same height as those of the silicon sample, they appear to the heated cantilever to be lower by ∼ 9 nm, resulting in a topographic distortion. Subtracting the topography obtained with the cold and hot cantilever yields, therefore, a map of the thermal conductivity features cross the sample.

Figure 18.12 shows the slopes associated with the components of the bending shown in Figure 18.10. One observes that the slope at $\eta = 0.5$, for the unheated and heated cantilever, are indeed maintained by the EFB for both the silicon sample and its oxide features.

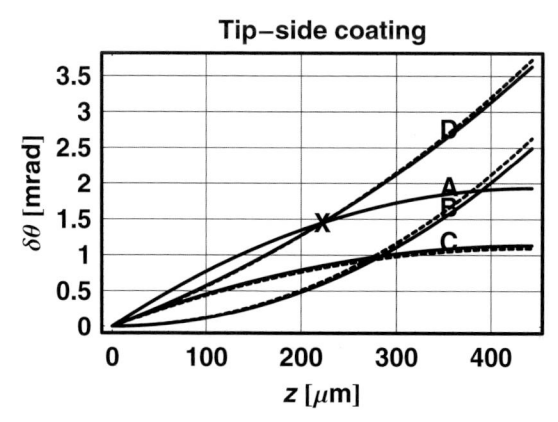

Fig. 18.12 The slope of the unheated (A) and heated (D) tip-side coated cantilever with the same notations as those of Figure 18.10. The solution of the double boundary condition problem indeed positions the slope at the point where the monitoring laser beam is incident on the cantilever, denoted by X, at the same value for the unheated and heated cantilever for both the silicon sample and its oxide features.

18.4.2
Top-Side Coating, Downward Thermal Bending: Si and SiO$_2$

Similarly, Figure 18.13 describes the different contributions to the bending of a top-side coated cantilever, where $\eta = 0.5$, with the same notations as those of Figure 18.10.

Figure 18.14 is an expanded view of the area denoted by O in Figure 18.13. As before, the proper solution of the double boundary condition problem positions the tip, denoted by V, for the unheated and heated cantilever, and for both the silicon and silicon oxide samples at $\delta y = 0$. Note that in contrast

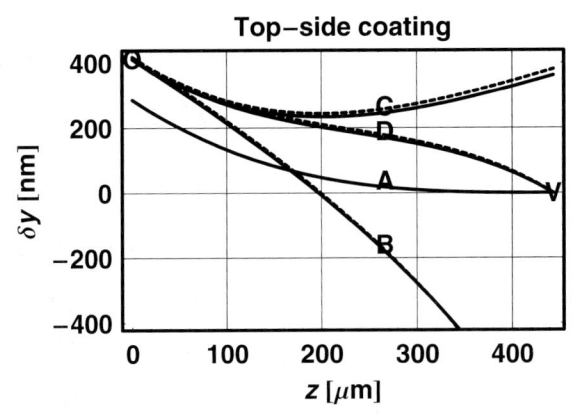

Fig. 18.13 The bending of the unheated and heated top-side coated cantilever with the same notations as those of Figure 18.10. As before, the solution of the double boundary condition problem positions the tip, denoted by V, for the unheated and heated cantilever and for both the silicon and silicon oxide samples at $\delta y = 0$.

to the tip-side coated cantilever, the oxide features now appear to be higher, rather than lower, although by the same amount of ~ 9 nm.

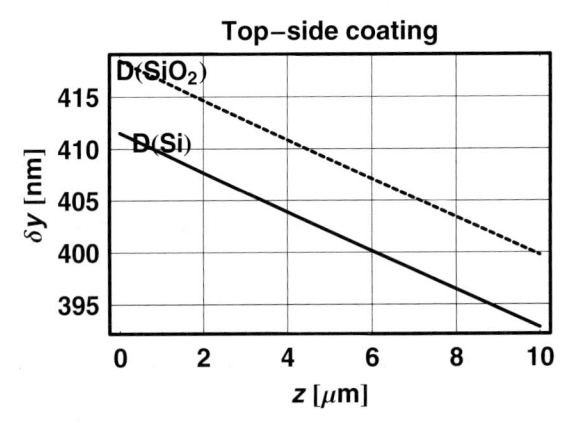

Fig. 18.14 An expanded view of the area denoted by O in Figure 18.13, where the total bending of the heated cantilever, up to 10 μm away from its base, is shown together with the position of the base of the cantilever for the silicon sample and its oxide features, D(Si) and D(SiO$_2$), respectively. Note that the tip is in contact with the sample at $\delta y = 0$.

Figure 18.15 shows the slopes associated with the components of the bending shown in Figure 18.13. One observes that the slope at $\eta = 0.5$ for the unheated and heated cantilever are indeed maintained by the EFB for both the silicon sample and its oxide features.

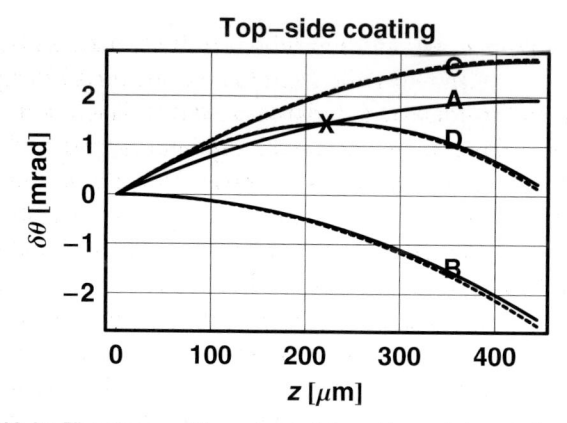

Fig. 18.15 The slopes of the unheated, top-side, metal-coated cantilever (A), the thermally induced (B) and mechanically induced (C) slope components of the heated cantilever and their sum, (D). The solid and dashed lines refer to the silicon sample and its oxide features, respectively. Note that the slope at η, denoted by X, is maintained for the unheated and heated cantilever.

18.4.3
Tip- and Top-Side Coating η-Dependent Apparent Height

Figure 18.16 shows the apparent height of the oxide features relative to the silicon sample, $\Delta_{Si} - \Delta_{SiO_2}$, obtained with a heated cantilever, as a function of η. The tip-side coated cantilever (solid line) yields an apparent height difference that is the negative of the one obtained with the top-side coated cantilever (dashed line). As noted before, the thermal bending of the cantilever is dominant near the tip while the mechanical bending is dominant next to its base. Thus there will be a point at which these two cancel each other out, and a change in the thermal bending will be compensated for exactly by an equal and opposite change in the mechanical bending. This value is dictated by the interplay of the mechanically and thermally induced slopes of a given cantilever.

18.4.4
Tip- and Top-Side Coating κ-Dependent Apparent Height

Finally, Figure 18.17 shows the apparent height difference, obtained with a cold and hot cantilever, as a function of the thermal conductivity of a sample, κ_s, relative to a sample with $\kappa_s = 1$. One observes that the response of the κ-AFM is most sensitive at lower thermal conductivities, where variations in the thermal conductivity will have a most pronounced effect on the cantilever temperature. However, throughout the range of conductivities shown, which

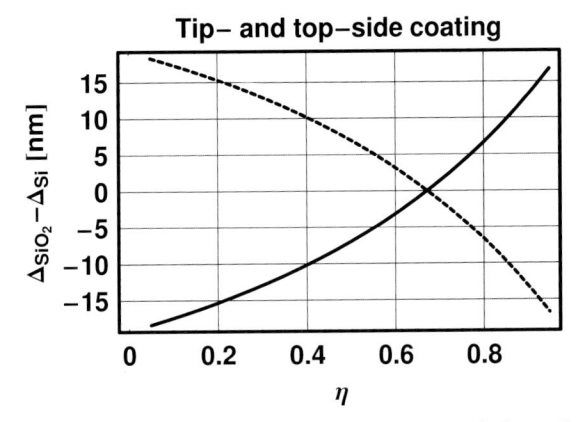

Fig. 18.16 The apparent height of the oxide features relative to the silicon sample, $\Delta_{SiO_2} - \Delta_{Si}$, obtained with a heated cantilever, as a function of η. The tip-side coated cantilever (solid line) yields an apparent height difference that is the negative of the one obtained with the top-side coated cantilever (dashed line).

encompass most materials of interest, the response is sufficient to ensure good differentiation between regions of differing thermal conductivities.

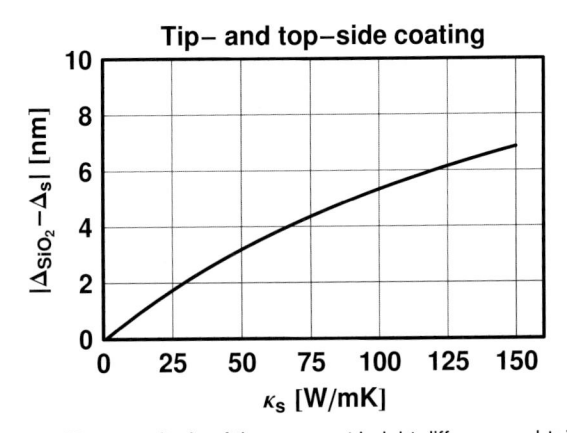

Fig. 18.17 The magnitude of the apparent height difference, obtained with a cold and hot cantilever, as a function of the thermal conductivity of a sample, κ_s, relative to a sample with $\kappa_s = 1$.

▶ **Exercises for Chapter 18**

1. Change the parameters of the rectangular cantilever, tip, and metal film and analyze the results obtained in this chapter.

2. Repeat the calculation using different values of ℓ_{ph} in Eq. (18.3) and Eq. (18.4) and discuss the results.

References

1. M. Nonnenmacher and H. K. Wickramasinghe, "Scanning probe microscopy of thermal conductivity and subsurface properties," Appl. Phys. Lett. **61**, 168 (1992).
2. O. Nakabeppu, M. Chandrachood, Y. Wu, J. Lai, and A. Majumdar, "Scanning thermal imaging microscopy using composite cantilever probes," Appl. Phys. Lett. **66**, 694 (1995).
3. D. G. Cahill, W. K. Ford, K. E. Goodson, G. D. Mahan, A. Majumdar, H. J. Maris, R. Merlin, and S. R. Phillpot, "Nanoscale thermal transport," J. Appl. Phys. **93**, 793 (2003).
4. E. T. Swartz, "Thermal boundary resistance," Rev. Mod. Phys. **61**, 605 (1989).
5. R. Prasher, "Predicting the thermal resistance of nanosized constrictions," Nano. Lett. **5**, 2155 (2005).
6. D. Sarid, B. Mc Carthy, R. Grover, "Scanning thermal conductivity microscope," Rev. Sci. Instrum. **77**, 023703 (2006).

19
Kelvin Probe Force Microscopy

Highlights

1. Tip–sample and cantilever–sample capacitance derivatives
2. Harmonic expansion of tip–sample electrostatic forces
3. Measurement of contact potential difference and its thermal limitation

Abstract

Models for tip–sample and cantilever–sample capacitance derivatives are presented together with numerical examples using typical parameters. The application of a tip–sample voltage consisting of both dc and ac components generates electrostatic forces at the first and second harmonics of the applied voltage. The harmonic expansion of these electrostatic forces shows that a tip–sample contact potential difference can be measured by tuning the applied dc voltage such that the force at the first harmonic vanishes. The lower bound to the measurement of the contact potential difference due to the thermal excitation of the cantilever is shown to be minimal.

19.1
Introduction

A Kelvin probe is an instrument that can measure the contact potential difference (CPD), $\Delta\Phi$, between the exposed surfaces of two samples of interest. Commercial Kelvin probes, which employ mm-scale electrodes, are too large for probing nm-scale structures. Kelvin probe force microscopy (KPFM), a derivative of atomic force microscopy (AFM), can provide CPD values with nm-scale spatial resolution. Kelvin probe and KPFM systems employ techniques that are similar in concept, yet vary subtly in the way they measure CPD. The key difference between these two techniques is that the Kelvin probe measures displacement currents while the KPFM measures tip–sample electrostatic forces. Specifically, the Kelvin probe employs two plane-parallel electrodes – the sample, which is stationary, and a reference electrode, which is vibrated with an amplitude $\Delta\delta$ and angular frequency ω. The inter-electrode modulated capacitance, C_{KP}, is given by

$$C_{KP} = \epsilon_0 \frac{A}{\delta(1 + \frac{\Delta\delta}{\delta}\sin\omega t)}, \tag{19.1}$$

where A is the area of each electrode, δ the sample–electrode mean spacing, and ϵ_0 the dielectric constant of free space. For $\Delta\delta/\delta \ll 1$, Eq. (19.1) reduces to

$$C_{KP} = \epsilon_0 \frac{A}{\delta}\left(1 - \frac{\Delta\delta}{\delta}\sin\omega t\right). \tag{19.2}$$

A dc voltage, v_{dc}, applied between these two electrodes, gives rise to a charge, Q, given by

$$Q = C_{KP}(\Delta\Phi - v_{dc}). \tag{19.3}$$

The displacement current, $i_{KP} = \partial Q/\partial t$, up to a phase, is given by

$$i_{KP} \propto \epsilon_0 \frac{A}{\delta^2}\omega(\Delta\Phi - v_{dc})\Delta\delta\sin\omega t. \tag{19.4}$$

In contrast, the KPFM, shown schematically in Figure 19.1, employs a conducting cantilever with a sharp tip at its end that is placed in close proximity to the sample. The application of dc and ac voltages, v_{dc} and v_{ac}, respectively, in the presence of the contact potential difference between the tip and the sample, results in an electrostatic force that vibrates the tip at the applied angular frequency, ω. We will show that the AFM photodiode which detects the vibration of the cantilever, generates a photocurrent, i_{KPFM}, given by

$$i_{KPFM} \propto \epsilon_0 \frac{r}{\delta}\frac{1}{k}(\Delta\Phi - v_{dc})v_{ac}\sin\omega t. \tag{19.5}$$

Here, r is the radius of the apex of the tip, and k the spring constant of the cantilever. For both systems, therefore, the external bias that nulls the current equals the contact potential difference. Note that in spite of the fact that the spatial distribution of the electric fields between the two electrodes differs markedly for the two systems, the value of the nulling external bias, $v_{dc} = \Delta\Phi$, should be the same. Note also that the sensitivities of a Kelvin Probe and of a KPFM are proportional to ω and $1/k$, respectively. Thus, high frequencies and soft cantilevers increase the sensitivities of these two systems, respectively. In the following we will analyze the operation of Kelvin probe force microscopy and give numerical examples that shed light on the magnitude of the relevant parameters.

The parameters describing the geometry of the AFM cantilever, given in Table 19.1, consist of its Young's modulus, mass density, length, width, thickness, and angle of inclination relative to the sample, E, ρ, L, w, d, and ϕ, respectively.

19 Kelvin Probe Force Microscopy

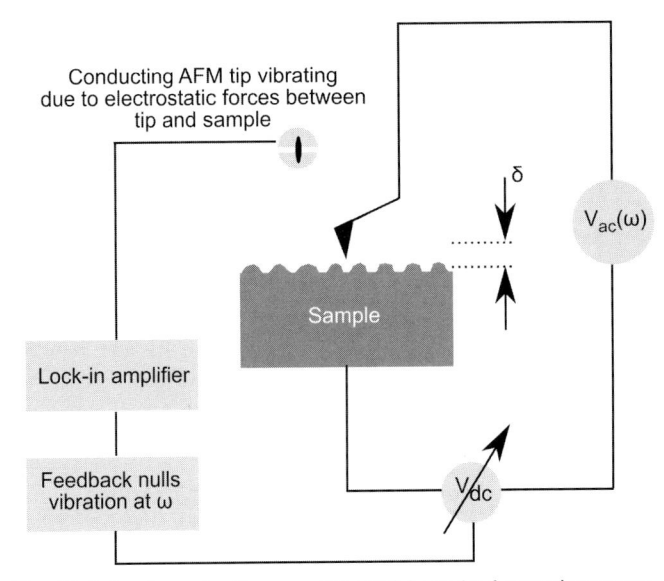

Fig. 19.1 A schematic diagram of the Kelvin probe force microscope.

Table 19.1 The cantilever's Young's modulus, mass density, length, width, thickness, and angle of inclination relative to the sample, E, ρ, L, w, d, and ϕ, respectively.

$E = 179\,\text{GN/m}^2$
$\rho = 2330\,\text{kg/m}^3$
$L = 125\,\mu\text{m}$
$w = 35\,\mu\text{m}$
$d = 1\,\mu\text{m}$
$\phi = 15\,\text{deg}$

The tip, attached to the free end of the cantilever, has a height h_t, apex radius r, and cone angle θ, given in Table 19.2.

Table 19.3 gives the resonance frequency and spring constant of the cantilever, f_0 and k, respectively, derived from Table 19.1, and the quality factor, Q, the system electronic bandwidth, B, and the tip–sample ac driving voltage, v_{ac}.

Table 19.2 The tip height h_t, apex radius r_t, and angle θ.

$h_t = 15\,\mu\text{m}$
$r_t = 40\,\text{nm}$
$\theta = 10\,\text{deg}$

Table 19.3 The resonance frequency and spring constant of the cantilever, f_0 and k, quality factor, Q, the system electronic bandwidth, B, and the tip–sample ac driving voltage, v_{ac}.

$f_0 = 89.27$ kHz
$k = 0.801$ N/m
$Q = 300$
$B = 1$ Hz
$v_{ac} = 2$ V

Finally, the Hamaker constant, A_H, and the chemical potential difference, $\Delta\Phi$, both associated with the tip–sample materials, are given in Table 19.4.

Table 19.4 The Hamaker constant, A_H, and the chemical potential difference, $\Delta\Phi$, associated with the tip–sample materials.

$A_H = 0.1$ aJ
$\Delta\Phi = 0.75$ V

19.2 Capacitance Derivatives

19.2.1 Tip–Sample Capacitance Derivative

Modeling the operation of the Kelvin probe force microscope involves the derivative of the capacitance of the probe–sample structure which consists of contributions from the cantilever and its tip. The following presents an accurate model for the contribution of the cantilever and three approximate models for the contribution of the tip. The simplest model for the tip–sample capacitance, C_t, replaces the tip by a sphere whose radius, r_t, equals that of the apex of the tip. C_t is given by the expansion

$$C_{t,a} = 4\pi\epsilon_0 r_t \sum_{n=1}^{\infty} \frac{\sinh(\alpha)}{\sinh(n\alpha)}, \tag{19.6}$$

where α is

$$\alpha = \ln\left(1 + \frac{\delta}{r_t} + \sqrt{\frac{\delta^2}{r_t^2} + 2\frac{\delta}{r_t}}\right), \tag{19.7}$$

and δ is the tip–sample spacing. The derivative of Eq. (19.6), $\partial C_t / \partial \delta$, is most conveniently performed numerically. However, one can obtain a rough approximation to this derivative using

$$\frac{\partial}{\partial \delta} C_t = 2\pi\epsilon_0 \frac{r_t}{\delta}. \tag{19.8}$$

In another model, the structure of the whole tip is taken as a cone whose apex forms part of a sphere. The resulting derivative of the capacitance of such a tip–sample structure, $\partial C_{t,b}/\partial \delta$, is

$$\frac{\partial}{\partial \delta} C_{t,b} = 2\pi\epsilon_0 \frac{r_t^2(1-\sin\theta)}{\delta(\delta+r_t(1-\sin\theta))} + \frac{\lambda^2}{2\pi\epsilon_0} \ln\left[\frac{(2\delta\sec\theta+r_t)^2}{4\delta\sec\theta(\delta\sec\theta+r_t)}\right], \tag{19.9}$$

where

$$\lambda = \frac{2\pi\epsilon_0}{\operatorname{arcsinh}(1/\tan\theta)}. \tag{19.10}$$

Here, θ is the half angle of the cone representing the tip and h_t the cone height. Examples of the three models that describe the tip–sample capacitance derivative, $\partial C_t/\partial \delta$, as a function of δ, are shown in Figure 19.2 for the range $r_t/10 < \delta < r$. In this figure, the solid, dashed, and dotted lines refer to $\partial C_t/\partial \delta$, $\partial C_{t,a}/\partial \delta$, and $\partial C_{t,b}/\partial \delta$, respectively. The three curves deviate from each other by up to a factor of 2. However, as will be shown later, their exact value drops out of the calculation of the performance of the Kelvin probe. We will, therefore, choose the simplest model, namely, that given by Eq. (19.8).

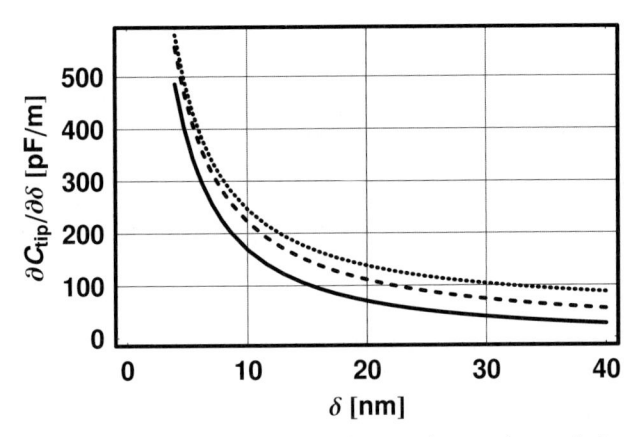

Fig. 19.2 Numerical examples of the tip–sample capacitance derivative, $\partial C_t/\partial \delta$, as a function of δ using the three models given in the text. The solid, dashed, and dotted lines refer to the three models described in the text.

19.2.2
Cantilever–Sample Capacitance Derivative

The derivative of the cantilever–sample capacitance, C_c, is given by

$$\frac{\partial}{\partial \delta} C_c = \epsilon_0 \frac{wL \sin \phi}{\phi(h_t + \delta + L \sin \phi)(h_t + \delta)}, \tag{19.11}$$

where L and w are the cantilever length and width, respectively, and ϕ is the angle of inclination of the cantilever relative to the sample. The cantilever–sample capacitance will be much larger than that of the tip–sample, unless the tip is close enough to the sample. It is important, therefore, to find out the range of tip–sample spacing, δ, such that the latter will dominate. Figure 19.3 shows the capacitance derivative of the tip–sample (solid line) and cantilever–sample structures, using Eq. (19.8) and Eq. (19.11), respectively. The latter is observed to be practically constant for the chosen range of δ values. The figure also depicts the value of δ_x for which these two capacitance derivatives are equal.

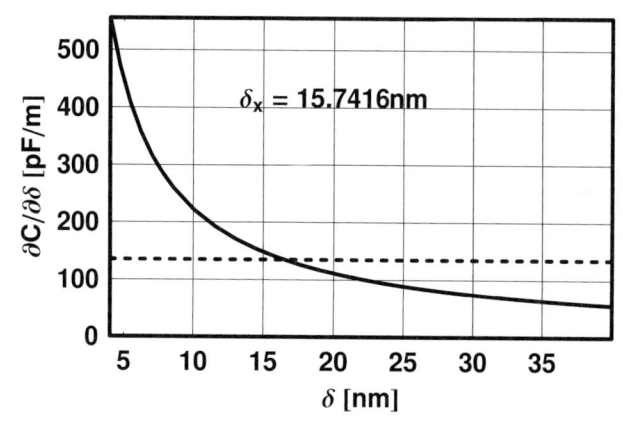

Fig. 19.3 The capacitance derivatives of the tip–sample (solid line) and cantilever–sample structures. The latter is practically constant for the chosen range of δ values. The figure also depicts the value of δ_x for which these two capacitance derivatives are equal.

19.3
Measurement of Contact Potential Difference

19.3.1
Tip–Sample and Cantilever–Sample Electrostatic Forces

The electrostatic energy, W_v, associated with a capacitor with capacitance, C, charged to a given voltage, v, is

$$W_v = \frac{1}{2}Cv^2, \tag{19.12}$$

and the electrostatic force derived from this energy, F_v, is

$$F_v = -\frac{\partial W}{\partial \delta}. \tag{19.13}$$

We will now use Eq. (19.8) as representing the tip–sample capacitance derivative, and consider the case where the force acts on the tip of a cantilever whose spring constant is k. Consider now an unbent cantilever whose tip is separated from a sample by a spacing δ, and apply a voltage, v, between the tip and a sample. The resulting electrostatic force will bend the cantilever, reducing δ to $\delta - F_v/k$. The force in this case can be obtained from the implicit function

$$F_v = \pi\epsilon_0 \frac{r_t}{\delta - F_v/k} v^2, \tag{19.14}$$

whose solution is

$$F_v = \frac{k\delta}{2}\left(1 - \sqrt{1 - 4\pi\epsilon_0 r_t \frac{v^2}{k\delta^2}}\right). \tag{19.15}$$

The critical voltage, v_c, at which the electrostatic force will overpower the cantilever bending force, is obtained by equating the square root in Eq. (19.10) to zero, yielding

$$v_c = \sqrt{\frac{k\delta^2}{4\pi\epsilon_0 r_t}}. \tag{19.16}$$

Applying a tip–sample voltage larger than v_c will therefore cause the tip to snap into the sample. The bending of the cantilever, Δ_c, given by

$$\Delta_c = \delta - \frac{F_v}{k} \tag{19.17}$$

is shown in Figure 19.4 as a function of v, where the solid, dashed, and dotted lines refer to cantilevers with k, $0.75k$, and $0.5k$, respectively.

For the case where

$$4\pi\epsilon_0 r_t \frac{v^2}{k\delta^2} \ll 1, \tag{19.18}$$

the electrostatic force can be approximated by

$$F_v = \pi\epsilon_0 \frac{r_t}{\delta} v^2. \tag{19.19}$$

It is of interest to compare the electrostatic force to the van der Waals force, F_{vdW}, acting between the tip and the sample,

$$F_{vdW} = -\frac{A_H r_t}{6\delta^2}. \tag{19.20}$$

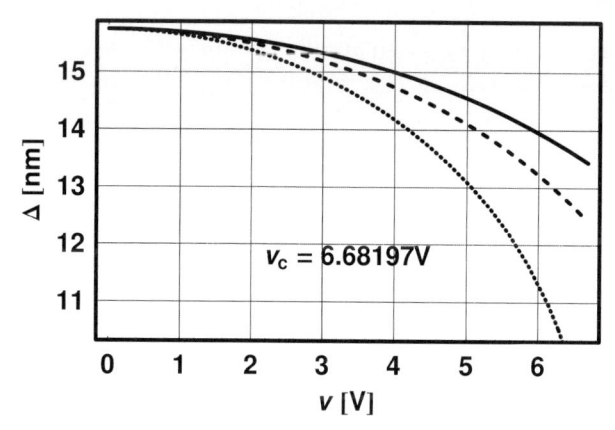

Fig. 19.4 The bending of the cantilever, Δ_c, as a function of v, for k (solid line), $0.75k$ (dashed line), and $0.5k$ (dotted line).

Figure 19.5 shows F_v (solid line) and F_{vdW} (dashed line) as a function of δ for $v = 1$ V. In a practical case, the van der Waals force can therefore be ignored unless one operates at too small spacings and too small voltages.

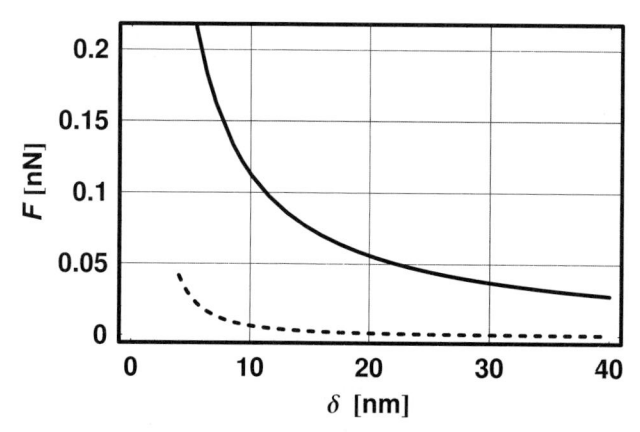

Fig. 19.5 F_v (solid line) and F_{vdW} (dashed line) as a function of δ for $v = 1$ V.

19.3.2
Harmonic Expansion of Tip–Sample Force

The operation of a Kelvin probe force microscope requires the application of both dc and ac voltages between the tip and the sample. The total voltage, given by

$$v = (\Delta\Phi - v_{dc}) + v_{ac}\sin(\omega t) \tag{19.21}$$

consists of a component v_{ac} at an angular frequency ω and a component v_{dc} which will be used to null $\Delta\phi$. Inserting Eq. (19.21) into Eq. (19.19) and using

$$\sin^2(\omega t) = \frac{1-\cos(2\omega t)}{2}, \tag{19.22}$$

yields the following three components of the force, F_v, one at dc,

$$F_{dc} = -\frac{1}{2}\frac{\partial C}{\partial \delta}\left[(\Delta\Phi - v_{dc})^2 + \frac{1}{2}v_{ac}^2\right], \tag{19.23}$$

a second one at ω

$$F_\omega = -\frac{\partial C}{\partial \delta}(\Delta\Phi - v_{dc})v_{ac}\sin(\omega t), \tag{19.24}$$

and a third one at 2ω,

$$F_{2\omega} = \frac{\partial C}{\partial \delta}\frac{1}{4}v_{ac}^2\cos(2\omega t). \tag{19.25}$$

The component at ω can be rewritten as

$$F_\omega = -\frac{\partial C}{\partial \delta}v_{ac,\,\text{effective}}\sin(\omega t), \tag{19.26}$$

where

$$v_{ac,\,\text{effective}} = (\Delta\Phi - v_{dc})v_{ac}, \tag{19.27}$$

is an effective ac voltage that can be nulled by tuning v_{dc}. We can now insert Eq. (19.8) into Eq. (19.23) to Eq. (19.25) and obtain the electrostatic deflection of the cantilever, Δ,

$$\Delta = \frac{F_v}{k}. \tag{19.28}$$

The resultant three deflection components are

$$\Delta_{dc} = -\frac{1}{2}\epsilon_0\pi\frac{r}{k\delta}\left[(\Delta\Phi - v_{dc})^2 + \frac{1}{2}v_{ac}^2\right] \tag{19.29}$$

at dc,

$$\Delta_\omega = -\epsilon_0\pi\frac{r}{k\delta}(\Delta\Phi - v_{dc})v_{ac}\sin(\omega t) \tag{19.30}$$

at ω, and

$$\Delta_{2\omega} = \frac{1}{4}\epsilon_0\pi\frac{r}{k\delta}v_{ac}^2\cos(2\omega t) \tag{19.31}$$

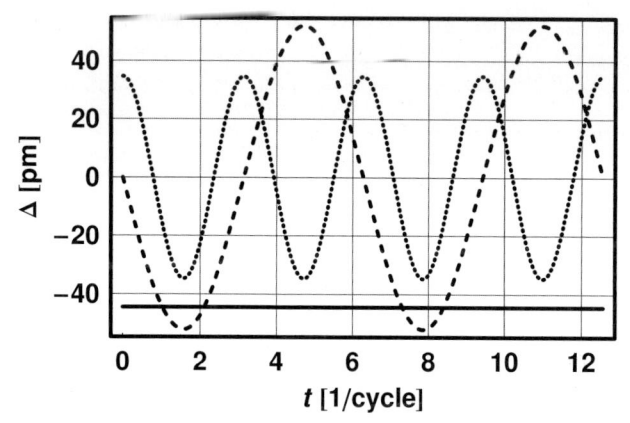

Fig. 19.6 The temporal response of the tip to the applied force, $\Delta_{c,dc}$ (solid line), $\Delta_{c,\omega}$ (dashed line), and $\Delta_{c,2\omega}$ (dotted line).

at 2ω. It is important to note that δ is a fixed tip–sample spacing that exists before the application of F_v. Figure 19.6 shows the temporal response of the tip of the cantilever to the applied force, Δ_{dc} (solid line), Δ_ω (dashed line), and $\Delta_{2\omega}$ (dotted line). Note that although the amplitude of vibration of the tip at ω is small, it can be readily detected using lock-in techniques.

19.3.3
Thermal Noise Limitations

It is useful to assess noise contributions when measuring low level signals, such as $\Delta_{c,\omega}$. To calculate the minimum observable value of $\Delta\Phi$, consider the rms amplitude of the thermal vibration of a cantilever, Δ_{th},

$$\Delta_{th} = \sqrt{\frac{2K_B TB}{\pi k Q f_0}}. \tag{19.32}$$

Equating Δ_ω to Δ_{th} in Eq. (19.30) leads to the minimum detectable value of $\Delta\Phi$, given by

$$\Delta\Phi_{min} = \frac{1}{v_{ac}\epsilon_0} \frac{\delta}{r_t} \sqrt{\frac{2K_B T k B}{\pi^3 Q f_0}}. \tag{19.33}$$

Table 19.5 gives the values of Δ_{th} and $\Delta\Phi_{min}$ for a tip–sample spacing $\delta = \delta_x$. One observes that thermal noise limitation of the contact potential difference, which is in the range $0.1\,\text{V} < \Delta\Phi < 1\,\text{V}$, can be made negligible by choosing proper parameters of the system, such as those given in Table 19.1 to Table 19.4.

Table 19.5 The values of Δ_{th} and $\Delta\Phi_{min}$ for $\delta = \delta_x$.

$\Delta_{th} = 0.011\,080\,1\,\text{pm}$
$\Delta\Phi_{min} = 0.062\,854\,2\,\text{mV}$

▶ **Exercises for Chapter 19**

1. Explore the results obtained in this chapter for several parameter ranges and discuss the numerical results.
2. Code the operation of a nonresonant dynamic force microscope as a Kelvin probe and explore the results for several parameter ranges.

References

1. M. P. O'Boyle, T. T. Hwang, and H. K. Wickramasinghe, "Atomic force microscopy of work functions on the nanometer scale," Appl. Phys. Lett. **74**, 2641 (1999).
2. G. Koley, M. G. Spencer, and H. R. Bhangale, "Cantilever effects on the measurement of electrostatic potentials by scanning kelvin probe microscopy," Appl. Phys. Lett. **79**, 545 (2001).
3. H. Yokoyama, T. Inoue, and J. Itoh, "Nonresonant detection of electric force gradients by dynamic force microscopy," Appl. Phys. Lett. **65**, 3143 (1994).
4. R. Grover, B. Mc Carthy, Y. Zhao, E. G. Jabbour, D. Sarid, G. M. Lows, B. R. Takulapalli, T. J. Thornton, and D. Gust, "Kelvin probe force microscopy as a tool for characterizing chemical sensors," Appl. Phys. Lett. **85**, 3926 (2004).
5. V. Palermo, M. Palma, and P. Samori, "Electronic characterization of organic thin films by Kelvin probe force microscopy," Adv. Mater. **18**, 145 (2006).

20
Raman Scattering in Nanocrystals

■ **Highlights**

1. Temperature dependence of Raman lineshape, linewidth, and frequency shift in a bulk silicon crystal
2. Temperature and size dependence of Raman lineshape, linewidth, and frequency shift in a silicon nanocrystal

Abstract

This chapter describes a model of temperature-dependent Raman scattering in a silicon crystal. The model is based on four-phonon anharmonic processes associated with the Raman active LO phonons of silicon. The model is presented in three stages that advance in their level of complexity. First, the temperature dependence of Stokes and anti-Stokes lineshape, linewidth and frequency shift of a silicon crystal are considered, and the ratio of their intensity discussed in terms of thermometry. Next, the model is extended to include size-dependent effects of spherical crystallites on room-temperature Raman scattering. Finally, the model is extended further to include also temperature-dependent effects. The final model can be used to analyze Raman scattering in nanostructures obtained by a variety of scanning probe microscopes.

20.1
Introduction

Inelastic scattering of light in a crystalline sample provides information about the optical modes of vibration at the center of its Brillouin zone. The scattered light, termed Raman scattering, consists of two spectra whose frequencies differ from that of the incident light, ω_0, as shown in Figure 20.1. The spectrum whose frequency is below that of the incident light, the Stokes component, has an intensity that is independent of the temperature of the sample. Its linewidth and line center, however, are temperature dependent due to four-phonon anharmonic processes associated with Raman-active LO phonons [1].

The spectrum whose frequency is above that of the incident light, the anti-Stokes component, is similar to that of the Stokes component, except that its intensity is also temperature dependent. Both the Stokes and anti-Stokes spectra in a crystal exhibit a Lorentzian shape, and the ratio of their intensity yields the temperature of the probed crystal. However, to use the ratio as a thermometer for an absorbing crystal such as silicon, one has to correct the measured ratio by taking into account the absorption of the incident light and the Stokes and anti-Stokes components, and their different Raman cross sections. The corrected ratio can then be used to accurately measure the temperature of a silicon crystal in the range of 150 K to above 1000 K. When the size of a spherical silicon crystal is reduced to below 30 nm, the lineshape of the Raman spectra starts to deviate from a Lorentzian shape. Studies of Raman scattering in polycrystalline silicon show that the reduction in size of a monocrystal exhibits a size-confinement effect that 1. produces an asymmetric lineshape, 2. broadens the linewidth, and 3. shifts the line center to lower frequencies. A model that accounts for these effects, taking place at room temperature, yields the spectra of a monocrystal as a function of its size [2]. An extension of this model that includes temperature effects was used to explain Raman spectra of cluster-deposited nanogranular silicon films [3]. Each one of these three models is presented together with numerical examples that shed light on the role of temperature and nanocrystal size.

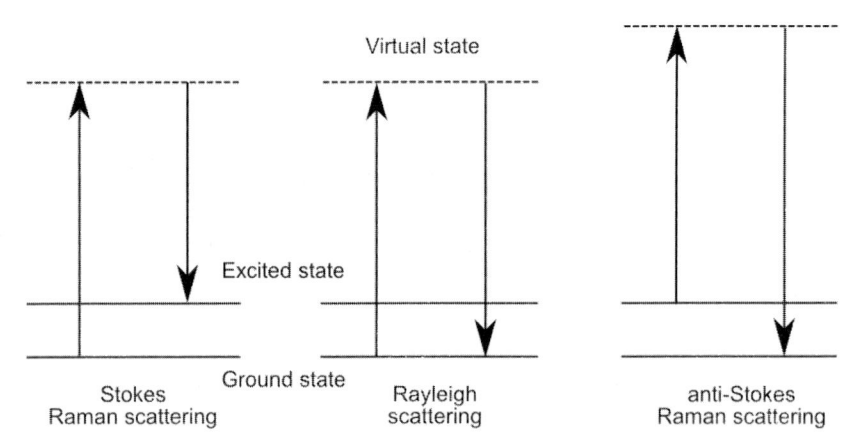

Fig. 20.1 A basic energy diagram of Raman scattering depicting the Stokes and anti-Stokes components.

20.2
Raman Scattering in Bulk Silicon Crystals as a Function of Temperature

20.2.1
Introduction

In this first section we treat an anharmonic model of Raman scattering in a silicon crystal whose size is larger than, say, 30 nm [1]. The range of frequencies, $\omega_{min} < \omega < \omega_{max}$, and temperatures, $T_1 < T < T_2$, used in the examples, are given in Table 20.1. Here and in the other two models presented in the next two sections, the frequencies are presented in units of cm^{-1}, which are converted in the code to units of Hz using the factor g.

Table 20.1 The conversion factor, g, and the range of frequencies, $\omega_{min} < \omega < \omega_{max}$, and temperatures, $T_1 < T < T_2$, used in the examples.

$$g = 188.365\,\text{GHz/cm}^{-1}$$
$$\omega_{min} = 460\,\text{cm}^{-1}$$
$$\omega_{max} = 540\,\text{cm}^{-1}$$
$$T_1 = 200\,\text{K}$$
$$T_2 = 1200\,\text{K}$$

20.2.2
Linewidth and Frequency Shift

A four-phonon Raman scattering process yields a Lorentzian-shaped spectrum for both the Stokes and anti-Stokes components. The temperature-dependent full-width at half maximum, Γ, of the Lorentzian is given by

$$\Gamma = A\left(1 + \frac{2}{x-1}\right) + B\left(1 + \frac{3}{y-1} + \frac{3}{(y-1)^2}\right), \tag{20.1}$$

where

$$x = e^{\frac{\hbar g \omega_0}{2K_B T}}, \tag{20.2}$$

and

$$y = e^{\frac{\hbar g \omega_0}{3K_B T}}. \tag{20.3}$$

The temperature-dependent frequency shift of the peak of the Lorentzian, describing both the Stokes and anti-Stokes components, Δ, is given by

$$\Delta = C\left(1 + \frac{2}{x-1}\right) + D\left(1 + \frac{3}{y-1} + \frac{3}{(y-1)^2}\right). \tag{20.4}$$

The total frequency shift of the Stokes and anti-Stokes, Ω, relative to the frequency of the incident light, is given by

$$\Omega = \omega_0 + \Delta. \tag{20.5}$$

Here, A, B, C, and D are the best-fit constants derived from experimental results. The examples for a silicon crystal, presented in this section, use the parameters given in Table 20.2.

Table 20.2 The experimental best-fit parameters used for the examples.

$\omega_0 = 528\,\text{cm}^{-1}$
$A = 1.295\,\text{cm}^{-1}$
$B = 0.105\,\text{cm}^{-1}$
$C = -2.96\,\text{cm}^{-1}$
$D = -0.174\,\text{cm}^{-1}$

The room temperature value of the functions x and y, the frequency shifts Δ and Ω, and the linewidth Γ are given in Table 20.3.

Table 20.3 The room temperature value of the functions x and y, the frequency shifts Δ and Ω, and the linewidth Γ.

$x = 6.680\,35$
$y = 3.547\,04$
$\Delta = -4.4616\,\text{cm}^{-1}$
$\Omega = 523.538\,\text{cm}^{-1}$
$\Gamma = 2.028\,19\,\text{cm}^{-1}$

Figure 20.2 shows (a) the frequency shift, Ω, and (b) the linewidth, Γ, as a function of temperature, T, both showing a dramatic change. Note that when the temperature increase, the linewidth naturally broadens and the frequency shift decreases due to softening of the acoustic modes of the crystal.

20.2.3
Spectra

The normalized lineshape of the Stokes, I_S, in terms of Γ and Ω, is given by

$$I_S = \frac{(\Gamma/2)^2}{(\omega - \Omega)^2 + (\Gamma/2)^2}. \tag{20.6}$$

The lineshape of the anti-Stokes, I_{aS}, in terms of I_S, is given by

$$I_{aS} = I_S e^{-\frac{\hbar g \omega_0}{K_B T}}, \tag{20.7}$$

20.2 Raman Scattering in Bulk Silicon Crystals as a Function of Temperature

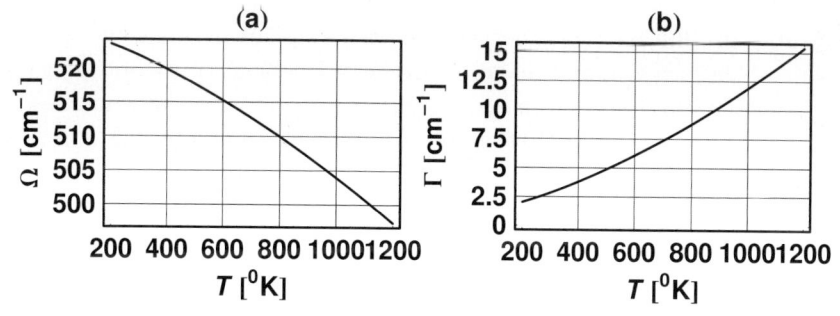

Fig. 20.2 (a) The frequency shift, Ω, and (b) the linewidth, Γ, as a function of the temperature, T.

where the temperature dependence is given by the Bose–Einstein occupation number. Figure 20.3 shows the Raman spectra at the temperatures of (a) 200 K and (b) 1200 K, where the solid and dashed lines refer to the Stokes and anti-Stokes components, respectively. Note that the Stokes components in both figures are normalized to unity, and that the anti-Stokes is frequency shifted so it is superimposed on the Stokes.

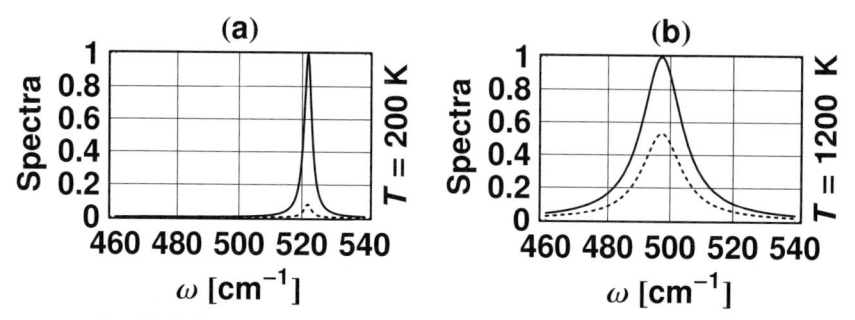

Fig. 20.3 Raman spectra at (a) 200 K and (b) 1200 K, where the solid and dashed lines refer to the Stokes and anti-Stokes components, respectively.

The experimentally observed ratio of the intensity of the anti-Stokes and Stokes components, $(I_{aS}/I_S)_{exp}$, has to be corrected for the different absorption coefficients and Raman cross sections associated with the three frequencies involved. The corrected ratio, that takes these effects into account, can be written as

$$\frac{I_S}{I_{aS}} = \frac{\alpha_I + \alpha_{aS}}{\alpha_I + \alpha_S} \left(\frac{\omega_S}{\omega_{aS}}\right)^3 \frac{S(\omega_0, \omega_S)}{S(\omega_0, \omega_{aS})} e^{\frac{\hbar g \omega_0}{K_B T}} \qquad (20.8)$$

where α_I, α_S, and α_{aS} are the absorption coefficients at the incident light and the Stokes and anti-Stokes components, respectively, and $S(\omega_0, \omega_S)$ and

$S(\omega_0, \omega_{aS})$ are the Raman cross sections of the Stokes and anti-Stokes components. Here, ω_0, ω_S, and ω_{aS} refer to the frequencies of the incident light and the Stokes and anti-Stokes components, respectively. The corrected experimental ratio should fit the theoretical one, given in Figure 20.4.

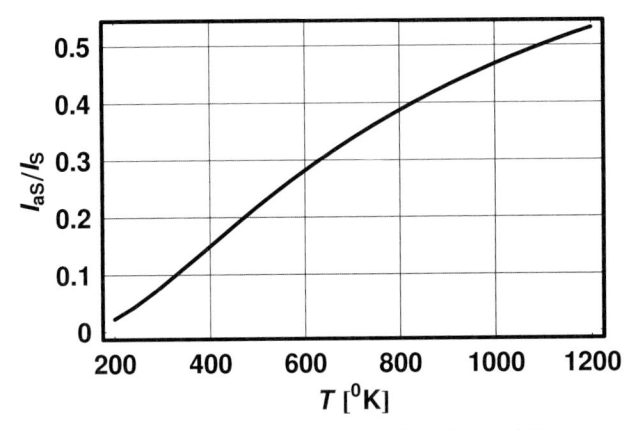

Fig. 20.4 The ratio of the intensity of the anti-Stokes and Stokes components, I_{aS}/I_S, as a function of the temperature, T.

20.3
Raman Spectra in Nanocrystals at Room Temperature

20.3.1
Introduction

This section extends the anharmonic model presented in the previous section by including spherical silicon crystals with a size-confinement effect, albeit at room temperature [2]. This model, which has been applied successfully to samples of polycrystalline silicon, will be accompanied by examples using frequencies in the range of $\omega_{min} < \omega < \omega_{max}$, given in Table 20.4.

Table 20.4 The range of frequencies, $\omega_{min} < \omega < \omega_{max}$, used in the examples.

ω_{min} =	$480\,\text{cm}^{-1}$
ω_{max} =	$540\,\text{cm}^{-1}$
$T =$	$300\,\text{K}$

20.3.2
Linewidth and Frequency Shift

The linewidth, Γ, in this model, is considered to be independent of the size, d, of the spherical silicon nanocrystal. The Stokes frequency shift, Ω, which does depend on d, is introduced through the phonon dispersion relation

$$\Omega = \sqrt{10^5[C_0 + \cos(\pi k/2)]}, \qquad (20.9)$$

where C_0 is an experimentally obtained best-fit constant, k is the phonon wave vector in units of $2\pi/a_0$, and a_0 is the silicon lattice constant. Table 20.5 gives the silicon lattice constant and the value of Γ and C_0. Also given is the value of the frequency shift for a large-size crystal where $k = 0$, which is denoted by $\Omega(0)$.

Table 20.5 The silicon lattice constant, a_0, the best-fit value of Γ and C_0, and the calculated value of $\Omega(0)$.

$$a_0 = 0.5483 \text{ nm}$$
$$\Gamma = 3.6 \text{ cm}^{-1}$$
$$C_0 = 1.714 \text{ cm}^{-1}$$
$$\Omega(0) = 520.961 \text{ cm}^{-1}$$

20.3.3
Spectra

The normalized lineshape of the Stokes component, given by I_S in terms of Γ, Ω, and d, is given by

$$I_S = \int_0^1 \frac{e^{-\left(k\frac{d}{2a_0}\right)^2}}{(\omega - \Omega)^2 + (\Gamma/2)^2} 4\pi k^2 \, dk, \qquad (20.10)$$

where the maximum value of k is 1. Figure 20.5 shows the normalized Stokes spectra in a nanocrystal with a size d (solid line) and in a bulk crystal (dashed line), both at 300 K. One observes that significant size-confinement effects exist in Raman scattering in a nanocrystal that exhibit (a) an asymmetric lineshape, (b) broadening of the linewidth, and (c) shifting the line center to lower frequencies.

The room temperature frequency shift in a bulk crystal, Ω_{bulk}, and in the nanocrystal, $\Omega_{nanocrystal}$, and their respective linewidth, Γ_{bulk} and $\Gamma_{nanocrystal}$, are given in Table 20.6, all of which are large enough to be measurable quantities.

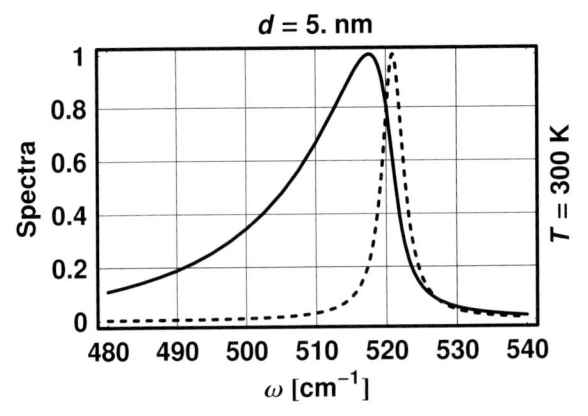

Fig. 20.5 The normalized Stokes spectra in a nanocrystal with a size d (solid line) and in a bulk crystal (dashed line), both at 300 K.

Table 20.6 The nanocrystal size, d, the room temperature frequency shift for a bulk crystal, Ω_{bulk}, and for a crystallite, $\Omega_{nanocrystal}$, and their respective linewidth, Γ_{bulk} and $\Gamma_{nanocrystal}$.

$$d = 5\,\text{nm}$$
$$\Omega_{bulk} = 520.97\,\text{cm}^{-1}$$
$$\Omega_{nanocrystal} = 517.68\,\text{cm}^{-1}$$
$$\Gamma_{bulk} = 3.6\,\text{cm}^{-1}$$
$$\Gamma_{nanocrystal} = 15.48\,\text{cm}^{-1}$$

20.4
Raman Spectra in Nanocrystals as a Function of Temperature

20.4.1
Introduction

The model presented in this section, which is a combination of the ones presented in the first two sections, enables one to calculate Raman scattering as a function of both the size of a spherical silicon nanocrystal and the temperature [3]. This model, which has been applied successfully to samples of cluster-deposited nanogranular silicon films, will be accompanied by examples using frequencies in the range $\omega_{min} < \omega < \omega_{max}$, and temperatures in the range $T_1 < T < T_2$, given in Table 20.7.

Table 20.7 The range of frequencies, $\omega_{min} < \omega < \omega_{max}$, and temperatures, $T_1 < T < T_2$, used in the examples.

$$\omega_{min} = 400 \text{ cm}^{-1}$$
$$\omega_{max} = 540 \text{ cm}^{-1}$$
$$T_1 = 300 \text{ K}$$
$$T_2 = 1200 \text{ K}$$

20.4.2 Linewidth and Frequency Shift

The modeling starts by considering the k-dependent phonon dispersion relation used in the previous section, which will now be denoted by Δ_k,

$$\Delta_k = \sqrt{10^5 [C_0 + \cos(\pi k/2)]}, \tag{20.11}$$

where k is the phonon wavevector in units of $2\pi/a_0$, and a_0 is the silicon lattice constant. The reason for this new notation is that in contrast to the previous model, here Δ_k is not the total frequency shift of the Raman spectrum but rather one of two contributions to the shift. Δ_k will now used to obtain the linewidth, Γ, which formally is similar to the one given in the first model, namely,

$$\Gamma = A\left(1 + \frac{2}{x-1}\right) + B\left(1 + \frac{3}{y-1} + \frac{3}{(y-1)^2}\right). \tag{20.12}$$

However, the functions x and y, which are now both temperature and size dependent, are given by

$$x = e^{\frac{\hbar g \Delta_k}{2K_B T}}, \tag{20.13}$$

and

$$y = e^{\frac{\hbar g \Delta_k}{3K_B T}}. \tag{20.14}$$

The second contribution to the frequency shift of the Raman spectrum, $\Delta_{k,T}$, is given by

$$\Delta_{k,T} = C\left(1 + \frac{2}{x-1}\right) + D\left(1 + \frac{3}{y-1} + \frac{3}{(y-1)^2}\right), \tag{20.15}$$

with the total shift, Ω, being

$$\Omega = \Delta_k + \Delta_{k,T}. \tag{20.16}$$

Table 20.8 gives the silicon lattice constant a_0 and the best-fit value of the parameters used in the examples.

Table 20.8 The silicon lattice constant, a_0, and the experimental best-fit parameters used in the examples.

$a_0 = 0.5483$ nm
$C_0 = 1.7$ cm^{-1}
$A = 1.683$ cm^{-1}
$B = 0.136$ cm^{-1}
$C = -3.996$ cm^{-1}
$D = -0.235$ cm^{-1}

Table 20.9 Room temperature value of the functions x and y.

$x = 3.47643$ cm^{-1}
$y = 2.29486$ cm^{-1}

The room temperature value of the functions x and y is given in Table 20.9.

20.4.3
Spectra

The normalized lineshape of the Stokes component, I_S, in terms of Γ, Ω, d, and T, is given by

$$I_S = \int_0^1 \frac{e^{-\left(k\frac{d}{2a}\right)^2}}{(\omega - \Omega)^2 + (\Gamma/2)^2} 4\pi k^2 \, dk, \qquad (20.17)$$

where the maximum value of k is 1. Figure 20.6 shows the normalized Raman Stokes spectra in a spherical silicon nanocrystal with a size d (solid line) and in a bulk crystal (dashed line), both at a temperature T denoted in the figure, and in a bulk silicon crystal at $T = 300$ (dotted line). In addition to the results obtained in the previous chapter, here both size confinement and temperature effects contribute to the Raman scattering in a nanocrystal that exhibit (a) an asymmetric lineshape, (b) broadening of the linewidth, and (c) shifting the line center to lower frequencies.

The nanocrystal size, d, the room temperature frequency shift in a bulk crystal, Ω_{bulk}, and in the nanocrystal, $\Omega_{\text{nanocrystal}}$, and their respective linewidth, Γ_{bulk} and $\Gamma_{\text{nanocrystal}}$ are given in Table 20.10, all large enough to be measurable quantities.

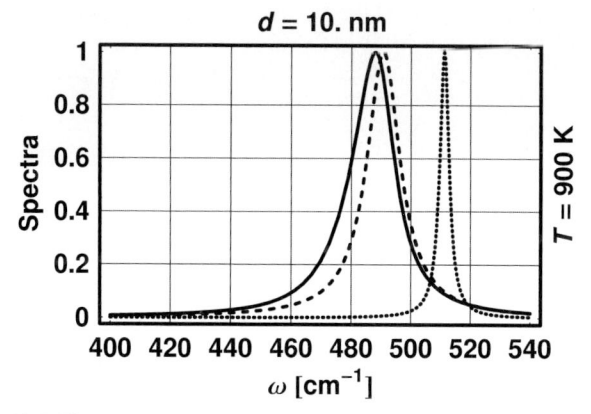

Fig. 20.6 The normalized Raman Stokes spectra in a nanocrystal with a size d (solid line) and in a bulk crystal (dashed line), both at a temperature T denoted in the figure, and in a bulk silicon crystal at $T = 300$ (dotted line).

Table 20.10 The spherical nanocrystal size, d, the Raman Stokes shift in a bulk crystal, Ω_{bulk}, and in a nanocrystal, $\Omega_{nanocrystal}$, and their respective linewidth, Γ_{bulk} and $\Gamma_{nanocrystal}$, both at 900 K.

$$d = 10\,\text{nm}$$
$$\Omega_{bulk} = 490.73\,\text{cm}^{-1}$$
$$\Omega_{nanocrystal} = 488.2\,\text{cm}^{-1}$$
$$\Gamma_{bulk} = 13.65\,\text{cm}^{-1}$$
$$\Gamma_{nanocrystal} = 16.81\,\text{cm}^{-1}$$

▶ **Exercises for Chapter 20**

1. Code and plot the linewidth of the Stokes component of the Raman spectra of a nanocrystal as a function of its size and as a function of temperature.
2. Code and plot the ratio of the anti-Stokes and Stokes components of the Raman spectra of a nanocrystal as a function of its size and as a function of temperature.
3. Code and plot the effect of the shape of a nanocrystal on its Raman spectra [4].

References

1. M. Balkanski, R. F. Wallis, and E. Haro, "Anharmomic effects in light scattering due to optical phonons in silicon," Phys. Rev. B **28**, 1928 (1983).
2. B. Pivac, K. Furić, and D. Desnica, "Raman line profile in polycrystalline silicon," J. Appl. Phys. **86**, 4383 (1999).
3. M. J. Konstantinović, S. Bersier, X. Wang, M. Hayne, P. Lievens, R. E. Silverans, and V. V. Moshehalkov, "Raman scattering in cluster-deposited nanogranular silicon films," Phys. Rev. B **66**, 161311–1 (2002)
4. I. H. Campbell and P. M. Fauchet, "The effects of microcrystal size and shape on the one phonon Raman spectra of crystalline semiconductors," Solid State Comm. **58**, 739 (1986).

Index

a

angular spring constant *see* cantilevers, angular spring constant
apparent barrier height *see* metal–insulator–metal tunneling, apparent barrier height
applied field *see* Fowler–Nordheim tunneling, applied field
average barrier height *see* metal–insulator–metal tunneling, average barrier height
average power *see* free vibrations, average power
averaged oscillations *see* Fowler–Nordheim tunneling, averaged oscillations

b

barrier height with image potential *see* metal–insulator–metal tunneling, barrier height with image potential
barrier width *see* metal–insulator–metal tunneling, barrier width
bending *see* cantilevers, bending
blockade *see* coulomb blockade, blockade
boundary resistance 256
buckling *see* cantilevers, buckling
bulk crystals 295

c

cantilevers
– angular spring constant 31, 36, 41, 63, 71
– bending 30, 66
– buckling 36, 62, 70
– characteristic functions 44, 76
– circular 49
– conversion 48
– displacement 32, 37, 59, 63, 67, 71
– linear spring constant 32, 37, 59, 60, 64, 68
– rectangular 27
– resonance frequencies 42, 74
– slope 31, 36, 41, 59, 63, 67, 71
– square 50
– triangular 58, 62
– twisting 40
– V-shape 56
– vibrations 42, 74
capacitance *see* electrostatics, capacitance
capacitance derivatives *see* Kelvin probe force microscopy, capacitance derivatives

characteristic functions see cantilevers, characteristic functions
charges inside the sphere see electrostatics, charges inside the sphere
charges outside the sphere see electrostatics, charges outside the sphere
circular see cantilevers, circular
confined structures see density of states, density of states in confined structures
constriction and boundary resistance
– combined electrical resistance 254
– constriction resistance 242, 248
– electronic density 246
– electronic density of states 247
– Maxwell limit 248
– Sharvin limit 251
constriction and bounday resistance
– boundary resistance 256
– electrical boundary resistance 260
– electronic specific heat 248
– Fermi k-vector 246
– Fermi temperature 245
– Fermi velocity 244
– free electron gas 244
– general media 256
– Lorenz number 244
– mean free path 246
– ratio of ℓ/σ 247
– thermal boundary resistance 256
– Wiedemann–Franz law 244
constriction resistance 248
contact force see tip–sample, contact force
contact potential difference 282

contact pressure see tip–sample, contact pressure
contact radius see tip–sample, contact radius
conversion see cantilevers, conversion
coulomb blockade
– blockade 179
– capacitance 180
– electrostatic energy 185
– quantum considerations 182
– quantum dot 185
– requirements and approximations 184
– staircase 184
– temperature effects 192
– tunneling current 186
– tunneling rates 186
coulomb staircase see coulomb blockade, staircase
critical points see density of states, critical points
cubical quantum dots see density of states, cubical quantum dots
current density see Fowler–Nordheim tunneling, current density

d
density of states 198
– critical points 207
– cubical quantum dots 204
– density of states in confined structures 203
– intraband optical transitions 207
– quantum wells 203
– quantum wires 204
– sphere in arbitrary dimensions 199
– spherical quantum dots 205

displacement   see cantilevers, displacement
downward thermal bending   see scanning thermal conductivity microscopy, downward thermal bending

e

effective tunneling area   see Fowler–Nordheim tunneling, effective tunneling area
electric field   see near field optics, electric field
electrical and thermal circuits   see scanning thermal conductivity microscopy, electrical and thermal circuits
electrical boundary resistance   260
electronic density   see constriction and boundary resistance, electronic density
electronic density of states   see constriction and boundary resistance, electronic density of states
electronic specific heat   see constriction and boundary resistance, electronic specific heat
electrostatic energy   see coulomb blockade, electrostatic energy
electrostatic force   see electrostatics, electrostatic force
electrostatics
– charges outside the sphere   216
– electrostatic force   220
– isolated point charge   212
– isolated sphere   214
– point charge and plane   212
– point charge and sphere   213
– potential and field   217
– sphere–plane   218
– two spheres   218

f

far-field solution   see near field optics, far-field solution
Feenstra's parameter   see scanning tunneling spectroscopy, Feenstra's parameter
Fermi k-vector   see constriction and bounday resistance, Fermi k-vector
Fermi temperature   see constriction and bounday resistance, Fermi temperature
Fermi velocity   see constriction and bounday resistance, Fermi velocity
Fermi–Dirac statistics   see scanning tunneling spectroscopy, Fermi–Dirac statistics
force curve   see tip–sample, force curve
Fowler–Nordheim tunneling
– applied field   164
– averaged oscillations   167
– current density   161
– effective tunneling area   168
– oscillation area   165
– oxide field   164
free electron gas   see constriction and bounday resistance, free electron gas
free vibrations
– analytical solution   112
– average power   117
– equation of motion   112, 117
– numerical solution   117
– phase   114
– Q dependent resonance   115
– steady state   119
– steepest slope   117
frequency shift   see Raman scattering, frequency shift

h

Hamaker constant   *see* tip–sample, Hamaker constant
hysterisis   *see* tip–sample, hysterisis

i

image potential   *see* metal–insulator–metal tunneling, image potential
indentation   *see* tip–sample, indentation
interaction   *see* tip–sample, interaction
intraband optical transitions   *see* density of states, intraband optical transitions
inverted functions   *see* tip–sample, inverted functions
isolated point charge   *see* electrostatics, isolated point charge

k

Kelvin probe force microscopy
– capacitance derivatives   285
– harmonic expansion of tip–sample force   289
– thermal noise limitations   291

l

large voltage approximation   *see* metal–insulator–metal tunneling, large voltage approximation
Lennard–Jones   *see* tip–sample, Lennard–Jones
linear spring constant   *see* cantilevers, linear spring constant
linewidth   *see* Raman scattering, linewidth
Lorenz number   *see* constriction and bounday resistance, Lorenz number

m

magnetic field   *see* near field optics, magnetic field
Maxwell limit   *see* constriction and boundary resistance, Maxwell limit
mean free path   *see* constriction and bounday resistance, mean free path
mechanical bending   *see* scanning thermal conductivity microscopy, mechanical bending
metal–insulator–metal tunneling
– apparent barrier height   157
– barrier height with image potential   152
– barrier width   153
– comparison of barriers   154
– general solution   148
– image potential   150
– large voltage approximation   149
– small voltage approximation   148
– tunneling current density   148

n

nanocrystals   *see* Raman scattering, nanocrystals
near field optics
– far-field solution   224
– intensity   237
– magnetic field   226
– near-field solution   229
– patterned apertures   238
– Poynting vector   227
– transformation   229
– vector potential   224
near field optics, electric field   224
near-field solution   *see* near field optics, near-field solution

noncontact mode 122

o

oscillation area *see* Fowler–Nordheim tunneling, oscillation area
oxide field *see* Fowler–Nordheim tunneling, oxide field

p

patterned apertures *see* near field optics, patterned apertures
phase diagram *see* tapping mode, phase diagram
phase shift *see* tapping mode, phase shift
point charge and plane *see* electrostatics, point charge and plane
point charge and sphere *see* electrostatics, point charge and sphere
potential and field *see* electrostatics, potential and field
Poynting vector *see* near field optics, Poynting vector
pull-out *see* tip–sample, pull-out
push-in *see* tip–sample, push-in

q

quantum consdirations *see* coulomb blockade, quantum considerations
quantum wells *see* density of states, quantum wells
quantum wires *see* density of states, quantum wires

r

Raman scattering
– frequency shift 295, 299, 301
– linewidth 295, 299, 301
– nanocrystals 298, 300
– spectra 296, 299, 300, 302
– temperature 300
ratio of ℓ/σ *see* constriction and bounday resistance, ratio of ℓ/σ
rectangular *see* cantilevers, rectangular
resonance frequencies *see* cantilevers, resonance frequencies

s

scanning thermal conductivity microscopy
– downward thermal bending 277
– electrical and thermal circuits 267
– mechanical bending 271
– thermal bending 271, 272
– thermal resistance 268–270
– thermal response 267
– upward thermal bending 275
scanning tunneling spectroscopy
– data processing 175
– Feenstra's parameter 173
– Fermi–Dirac statistics 172
– spectroscopy 174
– STS data file 175
Sharvin limit *see* constriction and boundary resistance, Sharvin limit
slope *see* cantilevers, slope
small voltage approximation *see* metal–insulator–metal tunneling, small voltage approximation
snap-in *see* tip–sample, snap-in
snap-out *see* tip–sample, snap-out
spectra *see* Raman scattering, spectra
sphere–plane *see* electrostatics, sphere–plane
square *see* cantilevers, square

steady state   *see* tapping mode, steady state
steepest slope   *see* free vibrations, steepest slope

t

tapping mode
– general solution   137
– indentation   136
– phase diagram   140
– phase shift   140
– steady state   139
– transient regime   138
thermal bending   *see* scanning thermal conductivity microscopy, thermal bending
thermal boundary resistance   256
thermal noise limitations   *see* Kelvin probe force microscopy, thermal noise limitations
thermal resistance   *see* scanning thermal conductivity microscopy, thermal resistance
thermal response   *see* scanning thermal conductivity microscopy, thermal response
tip–sample
– contact force   81, 84–86
– contact pressure   88
– contact radius   81, 83, 85, 86
– force curve   95, 96
– Hamaker constant   107
– hysterisis   101
– indentation   81, 83, 84, 91
– interaction   97
– inverted functions   85
– Lennard–Jones   97
– pull-out   93
– push-in   91
– snap-in   102
– snap-out   102
triangular   *see* cantilevers, triangular
tunneling current density   *see* metal–insulator–metal tunneling, tunneling current density
tunneling rates   *see* coulomb blockade, tunneling rates
twisting   *see* cantilevers, twisting

u

upward thermal bending   *see* scanning thermal conductivity microscopy, upward thermal bending

v

V-shape   *see* cantilevers, V-shape
vector potential   *see* near field optics, vector potential
vibrations   *see* cantilevers, vibrations

w

Wiedemann–Franz law   *see* constriction and bounday resistance, Wiedemann–Franz law